UNIVERSITY OF WINNIPEG
LIBRARY
515 Portage Avenue
Winnipeg, Manitoba R3B 2E9

TECHNOLOGY AND SOCIETY

TECHNOLOGY AND SOCIETY

Advisory Editor
DANIEL J. BOORSTIN, author of
The Americans and Director of
The National Museum of History
and Technology, Smithsonian Institution

THE STORY

OF THE

ATLANTIC TELEGRAPH

BY

HENRY M. FIELD

ARNO PRESS
A NEW YORK TIMES COMPANY
New York • 1972

Reprint Edition 1972 by Arno Press Inc.

Reprinted from a copy in The Newark Public Library

Technology and Society
ISBN for complete set: 0-405-04680-4
See last pages of this volume for titles.

Manufactured in the United States of America

Library of Congress Cataloging in Publication Data

   Field, Henry Martyn, 1822-1907.
      The story of the Atlantic telegraph.

     (Technology and society)
     First issued in 1866 under title: History of the Atlantic telegraph, to the return of the expedition of 1865.
     Reprint of the 1893 ed.
     1. Cables, Submarine--Atlantic. I. Title.
II. Series.
TK5611.F54 1972b   384.4'09182'1   72-5049
ISBN 0-405-04701-0

THE STORY

OF THE

ATLANTIC TELEGRAPH

BY

HENRY M. FIELD

"Since the discovery of Columbus, nothing has been done in any degree comparable to the vast enlargement which has thus been given to the sphere of human activity."
—THE TIMES, August 6th, 1858.

NEW YORK
CHARLES SCRIBNER'S SONS
1893

Copyright, 1892, by
HENRY M. FIELD.

Press of J. J. Little & Co.
Astor Place, New York

PREFACE

The recent death of Mr. Cyrus W. Field recalls attention to the great enterprise with which his name will be forever associated. "The Atlantic Telegraph," said the late Chief Justice Chase, "is the most wonderful achievement of civilization, and entitles its author to a distinguished rank among public benefactors. High upon that illustrious roll will his name be placed, and there will it remain while oceans divide, and telegraphs unite, mankind." The memory of such an achievement the world should not let die. The story of its varied fortunes reads like a tale of adventure. From the beginning it was a series of battles, fighting against the elements and against the unbelief of men. This long struggle the new generation may forget, profiting by the result, but thinking little of the means by which it was attained. What toil of hand and brain had gone before; what days and nights of watching and weariness; how often hope deferred had made the heart sick: how year after year had dragged on, and seen the end still afar off—all that is dimly remembered, even by those who reap the fruits of victory. And yet in the history of human achievements, it is necessary to trace these beginnings step by step, if we would learn the lesson they teach, that it is only out of heroic patience and perseverance that anything truly great is born.

Twelve years of unceasing toil was the price the Atlantic Telegraph cost its projector; and not years lighted up by the assurance of success, but that were often darkened with despair: years in which he was restlessly crossing and recrossing the ocean, only to find on either side, worse than storms and tempests, an incredulity which sneered at every failure, and derided the attempt as a delusion and a dream. Against such discouragements nothing could prevail but that faith, or fanaticism, which, believing the incredible, achieves the impossible. Such a tale, apart from the results, is in itself a lesson and an inspiration.

In attempting to chronicle all this, the relation of the writer to the prime mover has given him facilities for obtaining the materials of an authentic history; but he trusts that it will not lead him to overstep the limits of modesty. Standing by a new-made grave, he has no wish to indulge in undue praise even of the beloved dead. Enough for him is it to unroll the canvas on which the chief actor stands forth as the conspicuous figure. But in a work of such magnitude there are many actors, and there is glory enough for all; and it is a sacred duty to the dead to recognize, as he did, what was due to the brave companions in arms, who stood by him in disaster and defeat; who believed in him even when his own countrymen doubted and despaired; and furnished anew men and money and ships for the final conquest of the sea. If history records that the enterprise of the Atlantic Telegraph owed its inception to the faith and daring of an American, it will also record that all his ardor and activity would have been of no avail but for the science and seamanship, the capital and the undaunted courage, of England. But when all these conditions were supplied, it is the testimony of Englishmen themselves that his was the spirit within the wheels that made them revolve; that it was his intense vitality that infused itself into a great organization, and made the dream of science the reality of the world. This is not to his honor alone: it is a matter of national pride; and Americans may be pardoned if, in the year in which they celebrate the discovery of the continent, they recall that it was one of their countrymen whom the Great Commoner of England, John Bright, pronounced "the Columbus of our time, who, after no less than forty voyages across the Atlantic in pursuit of the great aim of his life, had at length by his cable moored the New World close alongside the Old." How the miracle was wrought, it is the design of these pages to tell.

CONTENTS

CHAPTER I.

DISCOVERY OF THE NEW WORLD BY COLUMBUS. RELATIVE POSITION OF THE TWO HEMISPHERES. NEAREST POINTS—THE OUTLYING ISLANDS, IRELAND AND NEWFOUNDLAND. SHORTER ROUTE TO EUROPE SUGGESTED BY BISHOP MULLOCK. THE ELECTRIC TELEGRAPH COMPANY OF NEWFOUNDLAND. PROJECT OF MR. F. N. GISBORNE. FAILURE OF THE COMPANY . . . . Page 1

CHAPTER II.

MR. GISBORNE COMES TO NEW YORK. IS INTRODUCED TO CYRUS W. FIELD, WHO CONCEIVES THE IDEA OF A TELEGRAPH ACROSS THE ATLANTIC OCEAN. IS IT PRACTICABLE? TWO ELEMENTS TO BE MASTERED, THE SEA AND THE ELECTRIC CURRENT. LETTERS OF LIEUTENANT MAURY AND PROFESSOR MORSE . . . . Page 15

CHAPTER III

MR. FIELD ENLISTS CAPITALISTS IN THE ENTERPRISE. COMMISSION TO NEWFOUNDLAND TO OBTAIN A CHARTER. THE NEW YORK, NEWFOUNDLAND, AND LONDON TELEGRAPH COMPANY . . . . Page 24

CHAPTER IV.

THE LAND-LINE IN NEWFOUNDLAND. FOUR HUNDRED MILES OF ROAD TO BE BUILT, A WORK OF TWO YEARS. ATTEMPT TO LAY A CABLE ACROSS THE GULF OF ST. LAWRENCE, IN 1855, FAILS. A SECOND ATTEMPT, IN 1856, IS SUCCESSFUL . . . . Page 38

CHAPTER V.

DEEP-SEA SOUNDINGS BY LIEUTENANT BERRYMAN IN THE DOLPHIN IN 1853, AND THE ARCTIC IN 1856, AND BY COMMANDER DAYMAN, OF THE BRITISH NAVY, IN THE CYCLOPS, IN 1857. THE BED OF THE ATLANTIC. THE TELEGRAPHIC PLATEAU . . . . Page 51

CONTENTS.

CHAPTER VI.

Mr. Field in London. The English Engineers and Electricians. Result of Experiments. The Atlantic Telegraph Company organized. Applies to the Government for Aid. Contract for a Cable. . . . . Page 69

CHAPTER VII.

Mr. Field returns to America. Seeks Aid from the Government. Opposition in Congress. Bill passed . . . . Page 91

CHAPTER VIII.

Return to England. The Niagara—Captain Hudson. The Agamemnon. Expedition of 1857 sails from Ireland. Speech of the Earl of Carlisle. The Cable broken . . . . . Page 112

CHAPTER IX.

Preparations for an Expedition in 1858. Mr. Field is made the General Manager of the Company. The Squadron assemble at Plymouth, and put to Sea, June 10. New Method of laying Cable, beginning in Mid-Ocean. The Agamemnon in Danger of being Foundered. The Cable lost Three Times. The Ships return to England. Meeting of the Directors. Shall they abandon the Project? One Last Effort . . . . . Page 142

CHAPTER X.

Second Expedition Successful. Cable landed in Ireland and Newfoundland . . . . Page 165

CHAPTER XI.

Great Excitement in America. Celebration in New York and other Cities . . . . Page 188

CHAPTER XII.

Sudden Stoppage of the Cable. Reaction of Public Feeling. Suspicions of Bad Faith. Did the Cable ever work? Page 213

CONTENTS.

CHAPTER XIII.

ATTEMPTS TO REVIVE THE COMPANY. THE GOVERNMENT ASKED FOR AID, BUT DECLINES TO GIVE AN UNCONDITIONAL GUARANTEE. FAILURE OF THE RED SEA TELEGRAPH. SCIENTIFIC EXPERIMENTS. CABLES LAID IN THE MEDITERRANEAN AND THE PERSIAN GULF. BRIEF HISTORY OF THE NEXT FIVE YEARS . . . . Page 229

CHAPTER XIV.

THE ENTERPRISE RENEWED. IMPROVEMENT ON THE OLD CABLE. THE GREAT EASTERN AND CAPTAIN ANDERSON. EXPEDITION OF 1865. TWELVE HUNDRED MILES LAID SAFELY, WHEN THE CABLE IS BROKEN . . . . Page 241

CHAPTER XV.

FORMATION OF A NEW COMPANY, THE ANGLO-AMERICAN. NEW CABLE MADE AND SHIPPED ON BOARD THE GREAT EASTERN . . Page 293

CHAPTER XVI.

THE EXPEDITION OF 1866. IMMENSE PREPARATIONS. RELIGIOUS SERVICE AT VALENTIA. SAILING OF THE FLEET. DIARY OF THE VOYAGE. CABLE LANDED AT HEART'S CONTENT . . . . Page 306

CHAPTER XVII.

RETURN TO MID-OCEAN TO SEARCH FOR THE CABLE LOST THE YEAR BEFORE. DRAGGING IN THE DEEP SEA. REPEATED FAILURES. CABLE FINALLY RECOVERED AND COMPLETED TO NEWFOUNDLAND . Page 347

CHAPTER XVIII.

THE AFTERGLOW. HONORS CONFERRED IN ENGLAND AND AMERICA. COMMERCIAL REVOLUTION WROUGHT BY THE CABLE. MR. FIELD AND THE ELEVATED RAILROADS IN NEW YORK CITY. TOUR ROUND THE WORLD. LAST YEARS. DEATH IN 1892. . . . . . Page 376

STORY OF THE ATLANTIC TELEGRAPH

CHAPTER I.

THE BARRIER OF THE SEA.

When Columbus sailed from the shores of Spain, it was not in search of a New World, but only to find a nearer path to the East. He sought a western passage to India. He had adopted a traditionary belief that the earth was round; but he did not once dream of another continent than the three which had been the ancient abodes of the human race—Europe, Asia, and Africa. All the rest was the great deep. The Florentine sage Toscanelli, from his knowledge of the world so far as then discovered, had made a chart, on which the eastern coast of Asia was represented as lying opposite to the western coast of both Europe and Africa. Accepting this theory, Columbus reasoned that he could sail direct from Spain to India. No intervening continent existed even in his imagination. Even after he had crossed the Atlantic, and descried the green woods of San Salvador rising out of the western seas, he thought he saw before him one of the islands of the Asiatic coast. Cuba he believed was a part of the mainland of India; Hayti was the Ophir of King Solomon; and when, on a later voyage, he came to the broad mouth of the Orinoco, and saw it pouring its mighty flood into the Atlantic, he rejoiced that he had found the great river Gihon, which had its rise in the garden of Eden! Even to the hour of his death, he remained ignorant of the real extent of his magnificent discovery. It was reserved to later times to lift the curtain fully from the world of waters; to reveal the true magnitude of the globe; and to unite the distant hemispheres by ties such as the great discoverer never knew.

It is hard to imagine the darkness and the terror which then hung over the face of the deep. The ocean to the west was a Mare Tenebrosum—a Sea of Darkness, into which only the boldest voyagers dared to venture. Columbus was the most successful navigator of his time. He had made voyages to the Western Islands, to Madeira and the Canaries, to Iceland on the north, and to the Portuguese settlements in Africa. But when he came to cross the sea, he had to grope his way almost blindly. But a few rays of knowledge glimmered, like stars, on the pathless waters. When he sailed on his voyage of discovery, he directed his course, first to the Canaries, which was a sort of outstation for the navigators of those times, as the last place at which they could take in supplies; and beyond which they were venturing into unknown seas. Here he turned to the west, though inclining southward toward the tropics (for even the great discoverers of that day, in their search for new realms to conquer, were not above the consideration of riches as well as honor, and somehow associated gems and gold with torrid climes), and bore away for India!

From this route taken by the great navigator, he crossed the ocean in its widest part. Had he, instead, followed the track of the Northmen, who crept around from Iceland to Greenland and Labrador; or had he sailed straight to the Azores, and then borne away to the north-west, he would much sooner have descried land from the mast-head. But steering in darkness, he crossed the Atlantic where it is broadest *and deepest;* where, as submarine explorers have since shown, it rolls over mountains, lofty as the Alps and the Himalayas, which lie buried beneath the surface of the deep. But farther north the two continents, so widely sundered, incline toward each other, as if inviting that closer relation and freer intercourse which the fulness of time was to bring.

As the island of Newfoundland is to stand in the foreground of our story, we observe on the map its salient geographical position. It holds the same relation to America that Ireland does to Europe. Stretching far out into the Atlantic, it is the vanguard of the western continent, or rather the signal-tower from which the New World may speak to the Old.

And yet, though large as England, and so near our

coast, few Americans ever see it, as it lies out of the track of European commerce. Our ships, though they skirt the Banks of Newfoundland, pass to the south, and get but occasional glimpses of the headlands. Even what is seen gives the country rather an ill reputation. It has a rockbound coast, around which hang perpetual fogs and mists, through which great icebergs drift slowly down, like huge phantoms of the deep, gliding away to be dissolved by the warm breath of the Gulf Stream : dangers that warn the voyager away from such a sea and shore.

Sailing west from Cape Race, and making the circuit of the island as far as the Straits of Belle Isle, one is often reminded of the most northern peninsula of Europe. The rocky shores are indented with numerous bays, reaching far up into the land, like the fiords along the coast of Norway; while the large herds of Caribou deer, that are seen feeding on the hills, might easily be mistaken for the flocks of reindeer that browse on the pastures and drink of the mountain torrents of ancient Scandinavia.

The interior of the island is little known. Not only is it uninhabited, it is almost unexplored, a boundless waste of rock and moor, where vast forests stretch out their unbroken solitudes, and the wild bird utters its lonely cry. Bears and wolves roam on the mountains. Especially common is the large and fierce black wolf; while of the smaller animals, whose skins furnish material for the fur-trade, such as martins and foxes, there is the greatest abundance. But from all pests of the serpent tribe, Newfoundland is as free as Ireland, which was delivered by the prayers of St. Patrick. There is not a snake or a frog or a toad in the island!

Yet, even in this ruggedness of nature, there is a wild beauty, which only needs to be " clothed upon " by the hand of man. Newfoundland, in many of its features, is not unlike Scotland, even in its most desolate portions, where the rocky surface of the country, covered with thick moss, reminds the emigrant Scot of the heather on his native moors. In the interior are lakes as long as Loch Lomond, and mountains as lofty as Ben Lomond and Ben Nevis. There are passes as wild as the Vale of Glencoe, where one might feel that he is in the heart of the Highlands, while the roar of the torrents yet more vividly recalls the

> Land of the brown heath and shaggy wood,
> Land of the mountain and the flood.

Yet in all this there is nothing to repel human habitation. By the hand of industry, these wild moors might be transformed into fruitful fields. We think it a very cold country, where winter reigns over half the year, as in Greenland; yet it is not so far north as Scotland, nor is its climate more inhospitable. It only needs the same population, the same hardy toil; and the same verdure would creep up its hill-sides, which now makes green and beautiful the loneliest of Scottish glens.

But at present the country is a *terra incognita*. In the interior there are no towns and no roads. As yet almost the whole wealth of the island is drawn from the sea. Its chief trade is its fisheries, and the only places of importance are a few small towns, chiefly on the eastern side, which have grown up around the trading posts. Besides these, the only settlements are the fishermen's huts scattered along the coast. Hence the bishop of the island, when he would make his annual visit to his scattered flock, is obliged to sail around his diocese in his yacht, since even on horseback it would not be possible to make his way through the dense forests to the remote parts of the island. This first suggested the idea of cutting across the island a nearer way, not only for internal intercourse, but for those who were passing to and fro on the sea.

It was in one of these excursions around the coast that the good Bishop Mullock, the head of the Roman Catholic Church in Newfoundland, when visiting the western portion of his diocese, lying one day becalmed in his yacht, in sight of Cape North, the extreme point of the province of Cape Breton, bethought himself how his poor neglected island might be benefited by being taken into the track of communication between Europe and America. He saw how nature had provided an easy approach to the mainland on the west. About sixty miles from Cape Ray stretched the long island of Cape Breton, while, as a stepping-stone, the little island of St. Paul's lay between. So much did it weigh upon his mind that, as soon as he got back to St. John's, he wrote a letter to one of the papers on the subject. As this was the first suggestion of a telegraph across Newfoundland, his letter is here given in full:

To the Editor of the Courier:

Sir: I regret to find that, in every plan for transatlantic communication, Halifax is always mentioned, and the natural capabilities of Newfoundland entirely overlooked. This has been deeply impressed on my mind by the communication I read in your paper of Saturday last, regarding telegraphic communication between England and Ireland, in which it is said that the nearest telegraphic station on the American side is Halifax, twenty-one hundred and fifty-five miles from the west of Ireland. Now would it not be well to call the attention of England and America to the extraordinary capabilities of St. John's, as the nearest telegraphic point? It is an Atlantic port, lying, I may say, in the track of the ocean steamers, and by establishing it as the American telegraphic station, news could be communicated to the whole American continent forty-eight hours, *at least*, sooner than by any other route. But how will this be accomplished? Just look at the map of Newfoundland and Cape Breton. From St. John's to Cape Ray there is no difficulty in establishing a line passing near Holy-Rood along the neck of land connecting Trinity and Placentia Bays, and thence in a direction due west to the Cape. You have then about forty-one to forty-five miles of sea to St. Paul's Island, with deep soundings of one hundred fathoms, so that the electric cable will be perfectly safe from icebergs. Thence to Cape North, in Cape Breton, is little more than twelve miles. Thus it is not only practicable to bring America two days nearer to Europe by this route, but should the telegraphic communication between England and Ireland, sixty-two miles, be realized, it presents not the least difficulty. Of course, we in Newfoundland will have nothing to do with the erection, working, and maintenance of the telegraph; but I suppose our Government will give every facility to the company, either English or American, who will undertake it, as it will be an incalculable advantage to this country. I hope the day is not far distant when St. John's will be the first link in the electric chain which will unite the Old World and the New. J. T. M.

St. John's, November 8, 1850.

This suggestion came at the right moment, since it quickened, if it did not originate, the first attempt to link the island of Newfoundland with the mainland of America. For about the same time, the attention of Mr. Frederick N. Gisborne, a telegraph operator, was attracted to a similar project. Being a man of great quickness of mind, he instantly saw the importance of such a work, and took hold of it with enthusiasm. It might easily occur to him without suggestion from any source. He had had much experience in telegraphs, and was then engaged in constructing a telegraph line in Nova Scotia. Whether, therefore, the idea was first

THE BARRIER OF THE SEA.   9

with him or with the bishop, is of little consequence. It might occur at the same time to two intelligent minds, and show the sagacity of both.

But having taken hold of this idea, Mr. Gisborne pursued it with indomitable resolution. As the labors of this gentleman were most important in the beginning of this work, it is a pleasure to recognize his untiring zeal and energy. In assurance of this we could have no higher authority than the following from the late Mr. E. M. Archibald, who was at the time Attorney-General of Newfoundland, and afterwards for many years British Consul at New York:

"It was during the winter of 1849-50, that Mr. Gisborne, who had been, as an engineer, engaged in extending the electric telegraph through Lower Canada and New Brunswick to Halifax, Nova Scotia, conceived the project of a telegraph to connect St. John's, the most easterly port of America, with the main continent. The importance of the geographical position of Newfoundland, in the event of a telegraph ever being carried across the Atlantic, was about the same time promulgated by Dr. Mullock, the Roman Catholic Bishop of Newfoundland, in a St. John's newspaper.

"In the spring of the following year (1851), Mr. Gisborne visited Newfoundland, appeared before the Legislature, then in session, and explained the details of his plan, which was an overland line from St. John's to Cape Ray, nearly four hundred miles in length, and (the submarine cable between Dover and Calais not having then been laid) a communication between Cape Ray and Cape Breton by steamer and carrier-pigeons, eventually, it was hoped, by a submarine cable across the Gulf of St. Lawrence. The Legislature encouraged the project, granted £500 sterling to enable Mr. Gisborne to make an exploratory survey of the proposed line to Cape Ray, and passed an act authorizing its construction, with certain privileges, and the appointment of commissioners for the purpose of carrying it out. Upon this, Mr. Gisborne, who was then the chief officer of the Nova Scotia Telegraph Company, returned to that province, resigned his situation, and devoted himself to the project of the Newfoundland telegraph. Having organized a local company for the purpose of constructing the first telegraph line in the island, from St. John's to Carbonear, a distance of sixty miles, he, on the fourth of September, set out upon the arduous expedition of a survey of the proposed line to Cape Ray, which occupied upward of three months, during which time himself and his party suffered severe privations, and narrowly escaped starvation, having to traverse the most rugged and hitherto unexplored part of the island.* On his return, having reported to the Legislature favorably of the project, and furnished estimates of the cost, he determined to proceed to New York, to obtain assistance to carry it out. . . . Mr. Gisborne returned to St. John's in the spring of 1852, when, at his instance, an act, incorporating himself (his being the only name mentioned in it) and such others as might become shareholders in a company, to be

---

* "On the fourth day of December, I accomplished the survey through three hundred and fifty miles of wood and wilderness. It was an arduous undertaking. My original party, consisting of six white men, were exchanged for four Indians; of the latter party, two deserted, one died a few days after my return, and the other, 'Joe Paul,' has ever since proclaimed himself an ailing man."—*Letter of Mr. Gisborne.*

called the Newfoundland Electric Telegraph Company, was passed, granting an exclusive right to erect telegraphs in Newfoundland for thirty years, with certain concessions of land, by way of encouragement, to be granted upon the completion of the telegraph from St. John's to Cape Ray. Mr. Gisborne then returned to New York, where he organized, under this charter, a company, of which Mr. Tebbets and Mr. Holbrook* were prominent members, made his financial arrangements with them, and proceeded to England to contract for the cable from Cape Ray to Prince Edward Island, and from thence to the mainland. Returning in the autumn, he proceeded in a small steamer, in November of that year, 1852, to stretch the first submarine cable, of any length, in America, across the Northumberland Strait from Prince Edward Island to New Brunswick, which cable, however, was shortly afterward broken, and a new one was subsequently laid down by the New York, Newfoundland, and London Telegraph Company. In the spring of the following year, 1853, Mr. Gisborne set vigorously to work to complete his favorite project of the line (which he intended should be chiefly underground) from St. John's to Cape Ray. He had constructed some thirty or forty miles of road, and was proceeding with every prospect of success, when, most unexpectedly, those of the company who were to furnish the needful funds dishonored his bills, and brought his operations to a sudden termination. He and the creditors of the company were for several months buoyed up with promises of forthcoming means from his New York allies, which promises were finally entirely unfulfilled ; and Gisborne, being the only ostensible party, was sued and prosecuted on all

---

* Horace B. Tebbets and Darius B. Holbrook.

sides, stripped of his whole property, and himself arrested to answer the claims of the creditors of the company. He cheerfully and honorably gave up every thing he possessed, and did his utmost to relieve the severe distress in which the poor laborers on the line had been involved."

This is a testimony most honorable to the engineer who first led the way through a pathless wilderness. But this Newfoundland scheme is not to be confounded with that of the Atlantic Telegraph, which did not come into existence until a year or two later. The latter was not at all included in the former. Indeed, Mr. Gisborne himself says, in a letter referring to his original project: "My plans were to run a subterranean line from Cape Race to Cape Ray, fly carrier-pigeons and run boats across the Straits of Northumberland to Cape Breton, and thence by overland lines convey the news to New York." He adds however: "Meanwhile Mr. Brett's experimental cable between Dover and Calais having proved successful, I set forth in my report, [which appeared a year after his first proposal], that 'carrier-pigeons and boats would be required only until such time as the experiments then making in England with submarine cables should warrant a similar attempt between Cape Ray and Cape Breton.'" But nowhere in his report does he allude to the possibility of ever spanning the mighty gulf of the Atlantic.

But several years after, when the temporary success

of the Atlantic Telegraph gave a name to everybody connected with it, he or his friends seemed not unwilling to have it supposed that this was embraced in the original scheme. When asked why he did not publish his large design to the world, he answered: " Because I was looked upon as a wild visionary by my friends, and pronounced a fool by my relatives for resigning a lucrative government appointment in favor of such a laborious speculation as the Newfoundland connection. Now had I coupled it at that time with an Atlantic line, all confidence in the prior undertaking would have been destroyed, and my object defeated." This may have been a reason for not announcing such a project to the public, but not for withholding it from his friends. A man can hardly lay claim to that which he holds in such absolute reserve.

However, whether he ever entertained the *idea* of such a project, is not a matter of the slightest consequence to the public, nor even to his own reputation. Ten years before Professor Morse had expressed, not a dreamer's fancy, but a deliberate conviction, founded on scientific experiments, that "a telegraphic communication might with certainty be established across the Atlantic Ocean;" so that the idea was not original with Mr. Gisborne, any more than with others who were eager to appropriate it.

It is a part of the history of great enterprises, that the moment one succeeds, a host spring up to claim the honor. Thus when, in 1858, the Atlantic Telegraph seemed to be a success, the public, knowing well who had borne the brunt and burden of the undertaking, awarded him the praise which he so well deserved; but instantly there were other Richmonds in the field. Those who had had no part in the labor, at least claimed to have originated the idea! Of course, these many claims destroy each other. But after all, to raise such a point at all is the merest trifling. The question is not who first had the "idea," but who took hold of the enterprise as a practical thing; who grappled with the gigantic difficulties of the undertaking, and fought the battle through to victory.

As to Mr. Gisborne, his activity in the beginning of the Newfoundland telegraph is a matter of history. In that preliminary work, he bore an honorable part, and acquired a title to respect, of which he cannot be deprived. All honor to him for his enterprise, his courage, and his perseverance!

But for the company of which he was the father, which he had got up with so much toil, it lived but a few months, when it became involved in debt some fifty thousand dollars, chiefly to laborers on the line, and ended its existence by an ignominious failure. The concern was bankrupt, and it was plain that, if the work was not to be finally abandoned, it must be taken up by stronger hands.

CHAPTER II.

CAN THE OCEAN BE SPANNED?

Mr. Gisborne left Halifax and came to New York in January, 1854. Here he took counsel with his friend Tebbets and others; but they could give him no relief. It was while in this state of suspense that he met, at the Astor House, Mr. Matthew D. Field, an engineer who had been engaged in building railroads and suspension bridges at the South and West. Mr. Field listened to his story with interest, and engaged to speak of it to his brother, Cyrus W. Field,* a merchant of New York, who had retired from business the year before, and had spent six months in travelling over the mountains of South America, from which he had lately returned. Accordingly, he introduced the subject, but found his brother disinclined to embark in any new undertaking. Though still a young man, his life had been for many years one of incessant devotion to business. He had accumulated an ample fortune,

---

* Born November 30, 1819, in Stockbridge, Massachusetts, the son of a Congregational minister, of whom three sons are still living: Mr. David Dudley Field, of New York; Mr. Justice Stephen J. Field, of the Supreme Court of the United States; and the writer of the present volume.

and was not disposed to renew the cares, the anxieties, and the fatigues of his former life. But listening to the details of a scheme which had in it much to excite interest, and which by its very difficulty stimulated the spirit of enterprise, he at length consented to see Mr. Gisborne, and invited him to his house. Accordingly he came, and spent an evening describing the route of his proposed telegraph, and the points it was to connect. After he left, Mr. Field took the globe which was standing in the library, and began to turn it over. *It was while thus studying the globe that the idea first occurred to him, that the telegraph might be carried further still, and be made to span the Atlantic Ocean.* The idea was not original with him, though he was to carry it out. It was indeed new *to him;* but it had long been a matter of speculation with scientific minds, though their theories had never attracted his attention. But once he had grasped the idea, it took strong hold of his imagination. Had the Newfoundland scheme stood alone, he would never have undertaken it. He cared little about shortening communication with Europe by a day or two, by relays of boats and carrier-pigeons. But it was the hope of further and grander results that inspired him to enter on a work of which no man could foresee the end.

An enterprise of such proportions, that would task to the utmost the science and the engineering skill of the world, was not to be rashly undertaken; and

CAN THE OCEAN BE SPANNED?

before giving a definite reply to Gisborne, Mr. Field determined to apply to the highest authorities in his own country.

The project of an Atlantic telegraph involved two problems: Could a cable be stretched across the ocean? and if it were, would it be good for anything to convey messages? The first was a question of mechanical difficulties, requiring a careful survey of the ocean itself, fathoming its depth, finding out the character of its bottom, whether level, or rough and volcanic; and all the obstacles that might be found in the winds that agitate the surface above, or the mighty currents that sweep through the waters below. The second problem was purely scientific, involving questions as to the laws of electricity, not then fully understood, and on which the boldest might feel that he was venturing on uncertain ground.

Such were the two elements or forces of nature to be encountered—the ocean and the electric current. Could they be controlled by any power of man? The very proposal was enough to stagger the faith even of an enthusiast. Who could lay a bridle on the neck of the sea? The attempt seemed as idle as that of Xerxes to bind it with chains. Was it possible to combat the fierceness of the winds and waves, and to stretch one long line from continent to continent? And then, after the work was achieved, would the lightning run along the ocean-bed from shore to shore? Such were the questions which had puzzled many an anxious brain, and which now troubled the one who was to undertake the work.

To get some light in his perplexity, Mr. Field, the very next morning after his interview with Gisborne, wrote two letters, one to Lieutenant Maury, then at the head of the National Observatory at Washington, on the nautical difficulties of the undertaking, asking if the sea were itself a barrier too great to be overcome; and the other to Professor Morse, inquiring if it would be possible to telegraph over a distance so great as that from Europe to America?

The mail soon brought an answer from Lieutenant Maury, which began: "Singularly enough, just as I received your letter, I was closing one to the Secretary of the Navy on the same subject." A copy of this he inclosed to Mr. Field, and it is given here. It shows the conclusions at which, even at that early day, scientific men were beginning to arrive:

"NATIONAL OBSERVATORY,
WASHINGTON, February 22, 1854.

"SIR: The United States brig Dolphin, Lieutenant Commanding O. H. Berryman, was employed last summer upon especial service connected with the researches that are carried on at this office concerning the winds and currents of the sea. Her observations were confined principally to that part of the ocean which the merchantmen, as they pass to and fro upon the business of trade between Europe and the

CAN THE OCEAN BE SPANNED? 19

United States, use as their great thoroughfare. Lieutenant Berryman availed himself of this opportunity to carry along also a line of deep-sea soundings, from the shores of Newfoundland to those of Ireland. The result is highly interesting, in so far as the bottom of the sea is concerned, upon the question of a submarine telegraph across the Atlantic; and I therefore beg leave to make it the subject of a special report.

"This line of deep-sea soundings seems to be decisive of the question of the practicability of a submarine telegraph between the two continents, *in so far as the bottom of the deep sea is concerned.* From Newfoundland to Ireland, the distance between the nearest points is about sixteen hundred miles ;* and the bottom of the sea between the two places is a plateau, which seems to have been placed there especially for the purpose of holding the wires of a submarine telegraph, and of keeping them out of harm's way. It is neither too deep nor too shallow ; yet it is so deep that the wires but once landed, will remain for ever beyond the reach of vessels' anchors, icebergs, and drifts of any kind, and so shallow, that the wires may be readily lodged upon the bottom. The depth of this plateau is quite regular, gradually increasing from the shores of Newfoundland to the depth of from fifteen hundred to two thousand fathoms, as you approach the other side. The distance between Ireland and Cape St. Charles, or Cape St. Lewis, in Labrador, is somewhat less than the distance from any point of Ireland to the nearest point of Newfoundland. But whether it would be better to lead the wires from Newfoundland or Labrador is not now

* From Cape Freels, Newfoundland, to Erris Head, Ireland, the distance is sixteen hundred and eleven miles ; from Cape Charles, or Cape St. Lewis, Labrador, to the same point, the distance is sixteen hundred and one miles.

the question; nor do I pretend to consider the question as to the possibility of finding *a time calm enough, the sea smooth enough, a wire long enough, a ship big enough,* to lay a coil of wire sixteen hundred miles in length; though I have no fear but that the enterprise and ingenuity of the age, whenever called on with these problems, will be ready with a satisfactory and practical solution of them.

"I simply address myself at this time to the question in so far as *the bottom of the sea* is concerned, and as far as that, the greatest practical difficulties will, I apprehend, be found after reaching soundings at either end of the line, and not in the deep sea. . . .

"A wire laid across from either of the above-named places on this side will pass to the north of the Grand Banks, and rest on that beautiful plateau to which I have alluded, where the waters of the sea appear to be as quiet and as completely at rest as at the bottom of a mill-pond. It is proper that the reasons should be stated for the inference that there are no perceptible currents, and no abrading agents at work at the bottom of the sea upon this telegraphic plateau. I derive this inference from a study of a physical fact, which I little deemed, when I sought it, had any such bearings.

"Lieutenant Berryman brought up with Brooke's deep-sea sounding apparatus specimens of the bottom from this plateau. I sent them to Professor Bailey, of West Point, for examination under his microscope. This he kindly gave, and that eminent microscopist was quite as much surprised to find, as I was to learn, that all those specimens of deep-sea soundings are filled with microscopic shells; to use his own words, *not a particle of sand or gravel exists in them.* These little shells, therefore, suggest the fact that there are

CAN THE OCEAN BE SPANNED? 21

no currents at the bottom of the sea whence they came; that Brooke's lead found them where they were deposited in their burial-place after having lived and died on the surface, and by gradually sinking were lodged on the bottom. Had there been currents at the bottom, these would have swept and abraded and mingled up with these microscopic remains the *débris* of the bottom of the sea, such as ooze, sand, gravel, and other matter; but not a particle of sand or gravel was found among them. Hence the inference that these depths of the sea are not disturbed either by waves or currents. Consequently, a telegraphic wire once laid there, there it would remain, as completely beyond the reach of accident as it would be if buried in air-tight cases. Therefore, so far as the bottom of the deep sea between Newfoundland, or the North Cape, at the mouth of the St. Lawrence, and Ireland, is concerned, the practicability of a submarine telegraph across the Atlantic is proved. . . .

"In this view of the subject, and for the purpose of hastening the completion of such a line, I take the liberty of suggesting for your consideration the propriety of an offer from the proper source, of a prize to the company through whose telegraphic wire the first message shall be passed across the Atlantic.

"I have the honor to be respectfully yours,

"M. F. MAURY,
"Lieutenant United States Navy.
"HON. J. C. DOBBIN, Secretary of the Navy."

The reply of Professor Morse showed equal interest in the subject, in proof of which he wrote that he would come down to New York to see Mr. Field about it. A few days after he came, and saw Mr. Field at his house. This was the beginning of an acquaintance which soon ripened into friendship, and which henceforth united these gentlemen together in this great achievement. Professor Morse, in conversation, entered at length into the laws of electricity as applied to the business of telegraphing, and concluded by declaring his entire faith in the undertaking as practical; as one that might, could, and would, be achieved. Indeed, this faith he had avowed years before. In a letter written as early as August tenth, 1843, to John C. Spencer, then Secretary of the Treasury, Professor Morse had detailed the results of certain experiments made in the harbor of New York to show the power of electricity to communicate at great distances, at the close of which he says—in words that now seem prophetic:

"The practical inference from this law is, that a telegraphic communication on the electro-magnetic plan may with certainty be established across the Atlantic Ocean! Startling as this may now seem, I am confident the time will come when this project will be realized."

It was the good fortune of Mr. Field—at that time and ever since—to have at hand an adviser in whose judgment he had implicit confidence. This was his eldest brother, David Dudley Field. They lived side by side on Gramercy Park, and were in daily communication. To the prudent counsels, wise judgment and unfaltering courage of the elder brother, the Atlantic Telegraph is more indebted than the world will ever know, for its first impulse and for the spirit which sustained it through long years of discouragement and disaster, when its friends were few. To this, his nearest and best counsellor, Mr. Field opened the project which had taken possession of his mind; and being strengthened by that maturer judgment, he finally resolved that, if he could get a sufficient number of capitalists to join him, he would embark in an enterprise which, beginning with the line to Newfoundland, involved in the end nothing less than an attempt to link this New World which Columbus had discovered, to that Old World which had been for ages the home of empire and of civilization. How the scheme advanced through the next twelve years, it will be our province to relate.

CHAPTER III.

THE COMPANY ORGANIZED.

AND so the young New York merchant set out to carry a telegraph across the Atlantic Ocean! The design had in it at least the merit of audacity. But whether the end was to be sublime or ridiculous time alone could tell. Certain it is that when his sanguine temper and youthful blood stirred him up to take hold of such an enterprise, he little dreamed of what it would involve. He thought lightly of a few thousands risked in an uncertain venture; but never imagined that he might yet be drawn on to stake upon its success the whole fortune he had accumulated; that he was to sacrifice all the peace and quiet he had hoped to enjoy; and that for twelve years he was to be almost without a home, crossing and re-crossing the sea, urging his enterprise in Europe and America. But so it is, that the Being who designs great things for human welfare, and would accomplish them by human instruments, does not lift at once the curtain from the stern realities they are to meet, nor reveal the rugged ascents they are to climb; so that it is only when at last the heights are attained, and they look backward, that they realize through what they have passed.

But could he find anybody to join him in his bold undertaking? Starving adventurers there always are, ready to embark in any Quixotic attempt, since they have nothing to lose. But would men of sense and of character; men who had fortunes to keep, and the habit which business gives of looking calmly and suspiciously at probabilities; be found to put capital in an enterprise where, if it failed, they would find their money literally at the bottom of the sea? It seemed doubtful, but he would try. His plan was, if possible, to enlist ten capitalists, all gentlemen of wealth, who together could lift a pretty heavy load; who, if need were, could easily raise a million of dollars, to carry out any undertaking.

The first man whom he addressed was his next-door neighbor, Mr. Peter Cooper, in whom he found the indisposition which a man of large fortune—now well advanced in life—would naturally feel to embark in new enterprises. The reluctance in this case was not so much to the risking of capital, as to having his mind occupied with the care which it would impose. These objections slowly yielded to other considerations. As they talked it over, the large heart of Mr. Cooper began to see that, if it were possible to accomplish such a work, it would be a great public benefit. This consideration prevailed, and what would not have been undertaken as a private speculation, was yielded to public interest. The conference ended by a conditional agreement to engage in it, if several others did, and, as we shall see, when the Company was organized, he became its President.

The early accession of this gentleman gave strength to the new enterprise. In all the million inhabitants of the city of New York there was not a name which was better known, or more justly held in honor, than that of Peter Cooper. A native of the city, where he had passed his whole life, he had seen its growth, from the small town it was after the War of the Revolution, and had himself grown with it. Beginning with very small means and limited opportunities, he had become one of its great capitalists. Many who thus rise to wealth, in the process of accumulation, form penurious habits which cling to them, and to the end of their days it is the chief object of life to hoard and to keep. But Mr. Cooper, while acquiring the fortune, had also the heart, of a prince; and used his wealth with a noble generosity. In the centre of New York stands to-day a massive building, erected at a cost of nearly a million of dollars, and consecrated " To Science and Art." This was Mr. Cooper's gift to his native city. Remembering his own limited advantages of education, he desired that the young men of

New York, the apprentices and mechanics, should have better opportunities than he had enjoyed. For this he endowed courses of lectures on the natural sciences; he opened the largest reading-room in America, which furnishes a pleasant resort to thousands of readers daily; while to help the other sex, he added a School of Design for Women, which trains hundreds to be teachers, and some of them artists; who go forth into the world to earn an honest living, and to bless the memory of their generous benefactor. This noble institution, standing in the heart of the city, is his enduring monument.

Yet while doing so much for the public, those who saw Peter Cooper in his family knew how he retained the simple habits of early life — how, while giving hundreds of thousands to others, he cared to spend little on himself; how he remained the same modest, kindly old man; the pure, the generous, and the good. His was

"The good gray head that all men knew,"

and that was sadly missed when, nearly thirty years after, in 1883, at the age of ninety-two, he was borne to his grave. It is a pleasant remembrance that the beginning of this enterprise was connected with that honored name.

Mr. Field next addressed himself to Mr. Moses Taylor, a well-known capitalist of New York, engaged in extensive business reaching to different parts of the world, and whose daily observation of all sorts of enterprises, both sound and visionary, made him perhaps a severer judge of any new scheme. With this gentleman he had then no personal acquaintance, but sent a note of introduction from his brother, David Dudley Field, with a line requesting an interview, to which Mr. Taylor replied by an invitation to his house on an evening when he should be disengaged. As these two gentlemen afterwards became very intimately associated, they often recurred to their first interview. Said Mr. Field : " I shall never forget how Mr. Taylor received me. He fixed on me his keen eye, as if he would look through me : and then, sitting down, he listened to me for nearly an hour without saying a word." This was rather an ominous beginning. However, his quick mind soon saw the possibilities of the enterprise, and the evening ended by an agreement—conditional, like Mr. Cooper's—to enter into it.

Mr. Taylor, being thus enlisted, brought in his friend, Mr. Marshall O. Roberts—a man whose career has been too remarkable to be passed without notice. A native of the City of New York, (though his father was a physician from Wales, who came to this country early in this century,) he found himself, when a boy of eight years, an orphan, without a friend in the world. From that time he made his way purely by his own industry and indomitable will. At the age of twenty he was embarked in business for himself, and his history soon became a succession of great enterprises. If we were to relate some of the incidents connected with his rise of fortune, they would sound more like romance than reality. He was the first to project those floating palaces which now ply the waters of the Hudson and the great lakes. He was one of the early promoters of the Erie Railroad. When the discovery of gold in California turned the tide of emigration to that coast, he started the line of steamers to the Isthmus of Panama, and controlled largely the commerce with the Pacific. Thus his hand was felt, giving impulse to many different enterprises on land and sea. His whole course was marked by a spirit of commercial daring, which men called rashness, until they saw its success, and then applauded as marvellous sagacity.

Mr. Field next wrote to Mr. Chandler White, a personal friend of many years' standing, who had retired from business, and was living a few miles below the city, near Fort Hamilton, at one of those beautiful points of view which command the whole harbor of New York. He too was very slow to yield to argument or persuasion. Why should he—when he had cast anchor in this peaceful spot—again embark in the cares of business, and, worst of all, in an enterprise the scene of which was far distant, and the results very uncertain? But enthusiasm is always magnetic, and the glowing descriptions of his persuader at length prevailed.*

There were now five gentlemen enlisted; and Mr. Field was about to apply to others, to make up his proposed number, when Mr. Cooper came to ask why *five* would not do as well as *ten ?* The question was no sooner asked than answered. To this all agreed, and at once fixed an evening when they should meet at Mr. Field's house to hear his statements and to examine the charter of the old company, find out what it had done, and what it proposed to do, what property it had and what debts it owed; and decide whether the enterprise offered sufficient inducements to embark in it. Accordingly they met, and for four nights in succession discussed the subject. It was in the dining-room of Mr. Field's house, and the large table was spread with maps of the route to be traversed by the line of telegraph, and with plans and estimates of the work to be done, the cost of doing it, and the return which they might hope in the end to realize for their labor and their capital. The result was an agreement

---

*Although it is anticipating a year in time, I cannot resist the pleasure of adding here the name of another eminent merchant, who afterward joined this little Company, Mr. Wilson G. Hunt. Mr. Hunt is one of the old merchants of New York who, through his whole career, has maintained the highest reputation for commercial integrity, and whose fortune is the reward of a long life of honorable industry. He joined the Company in 1855, and was a strong and steady friend through all its troubles till the final success.

THE COMPANY ORGANIZED. 31

on the part of all to enter on the undertaking, if the Government of Newfoundland would grant a new charter conceding more favorable terms. To secure this it was important to send at once a commission to Newfoundland. Neither Mr. Cooper, Mr. Taylor, nor Mr. Roberts could go; and it devolved on Mr. Field to make the first voyage on this business, as it did to make many voyages afterwards to Newfoundland, and still more across the Atlantic. But not wishing to take the whole responsibility, he was accompanied at his earnest request by Mr. White, and by Mr. D. D. Field, whose counsel, as he was to be the legal adviser of the Company, was all-important in the framing of the new charter that was to secure its rights. The latter thus describes this first expedition:

"The agreement with the Electric Telegraph Company, and the formal surrender of its charter, were signed on the tenth of March, [1854,] and on the fourteenth we left New York, accompanied by Mr. Gisborne. The next morning we took the steamer at Boston for Halifax, and thence, on the night of the eighteenth, departed in the little steamer Merlin for St. John's, Newfoundland. Three more disagreeable days, voyagers scarcely ever passed, than we spent in that smallest of steamers. It seemed as if all the storms of winter had been reserved for the first month of spring. A frost-bound coast, an icy sea, rain, hail, snow and tempest, were the greetings of the telegraph adventurers in their first movement toward Europe. In the darkest night, through which no man could see the ship's length, with snow filling the air and flying into the eyes of the sailors, with ice in the water, and a heavy sea rolling and moaning about us, the captain felt his way around Cape Race with his lead, as the blind man feels his way with his staff, but as confidently and as safely as if the sky had been clear and the sea calm; and the light of morning dawned upon deck and mast and spar, coated with glittering ice, but floating securely between the mountains which form the gates of the harbor of St. John's. In that busy and hospitable town, the first person to whom we were introduced was Mr. Edward M. Archibald, then Attorney-General of the Colony, and now British Consul in New York. He entered warmly into our views, and from that day to this, has been an efficient and consistent supporter of the undertaking. By him we were introduced to the Governor, (Kerr Bailey Hamilton,) who also took an earnest interest in our plans. He convoked the Council to receive us, and hear an explanation of our views and wishes. In a few hours after the conference, the answer of the Governor and Council was received, consenting to recommend to the Assembly a guarantee of the interest of £50,000 of bonds, an immediate grant of fifty square miles of land, a further grant to the same extent on the completion of the telegraph across the ocean, and a payment of £5,000 toward the construction of a bridle-path across the island, along the line of the land telegraph."

This was a hopeful beginning; and, though the charter was not yet obtained, feeling assured by this official encouragement, and the public interest in the project, that it would be granted by the colony, Mr. Field remained in St. John's but three days, when he took the Merlin back to Halifax on his way to

THE COMPANY ORGANIZED. 33

New York, there to purchase and send down a steamer for the service of the Company, leaving his associates to secure the charter and to carry out the arrangements with the former company. To settle all these details was necessarily a work of time. First, the charter of the old Electric Telegraph Company had to be repealed, to clear the way for a new charter to the Company, which was to bear the more comprehensive title of " New York, Newfoundland, *and London*." This charter—which had been drawn with the greatest care by the counsel of the Company, while on the voyage to Newfoundland—bore on its very front the declaration that the plans of the new Company were much broader than those of the old. In the former charter, the design was thus set forth :

"The telegraph line of this company is designed to be strictly an 'Inter-Continental Telegraph.' Its termini will be New York, in the United States, and London, in the kingdom of Great Britain ; these points are to be connected by a line of electric telegraph from New York to St. John's, Newfoundland, partly on poles, partly laid in the ground, and partly through the water, *and a line of the swiftest steamships ever built from that point to Ireland.* The trips of these steamships, it is expected, will not exceed five days, and as very little time will be occupied in transmitting messages between St. John's and New York, the communication between the latter city and London or Liverpool, will be effected *in six days*, or less. The company will have likewise stationed at St. John's a steam yacht, for the purpose of intercepting the European and American steamships, so that no opportunity may be lost in forwarding intelligence in advance of the ordinary channels of communication."

But the charter of the New York, Newfoundland, and London Telegraph Company, which was now to be obtained, began by declaring, in its very first sentence: "Whereas it is deemed advisable to establish a line of telegraphic communication between America and Europe by way of Newfoundland." Not a word is said of fast ships, of communications in less than six days, but every thing points to a line across the ocean. Thus one section gives authority to establish a submarine telegraph across the ocean, from Newfoundland to Ireland; another section prohibits any other company or person from touching the coast of Newfoundland or its dependencies [which includes Labrador] with a telegraphic cable or wire, from any point whatever, for fifty years; and a third section grants the Company fifty square miles of land upon the completion of the submarine line across the Atlantic.

In other respects the charter was equally liberal. It incorporated the associates for fifty years, established perfect equality, in respect to corporators and officers, between citizens of the United States and British subjects, and allowed the meetings of the stockholders and directors to be held in New York, in Newfoundland, or in London.

To obtain such concessions was a work of some diffi-

THE COMPANY ORGANIZED. 35

culty and delay. The Legislature of the province were naturally anxious to scan carefully conditions that were to bind them and their children for half a century. I have now before me the papers of St. John's of that day, containing the discussions in the Legislature; and while all testify to the deep public interest in the project, they show a due care for the interests of their own colony, which they were bound to protect. At length all difficulties were removed, and the charter was passed unanimously by the Assembly, and confirmed by the Council.

This happy result was duly celebrated, in the manner which all Englishmen approve, by a grand dinner given by the commissioners of the new Company, to the members of the Assembly and other dignitaries of the colony, at which there were eloquent prophecies of the good time coming, showing how heartily the enterprise was welcomed by all classes; and how fond were the anticipations of the increased intercourse it would bring, and the manifold benefits it would confer on their long-neglected island.

No sooner were the papers signed, than the wheels, so long blocked, were unloosed, and the machinery began to move. Mr. White at once drew on New York for fifty thousand dollars, and paid off all the debts of the old company. A St. John's newspaper of April 8th, 1854, amid a great deal on the subject, contains this paragraph, which is very significant of the dead state of the old company, and of the life of the new:

"The office of the new Electric Telegraph Company has been surrounded the last two or three days by the men who had been engaged the last year on the line, and who are being paid all debts, dues, and demands against the old association. We look upon the readiness with which these claims are liquidated as a substantial indication on the part of the new Company that they will complete to the letter all that they have declared to accomplish in this important undertaking."

In the early part of May, the two gentlemen who had remained behind in Newfoundland rejoined their associates in New York, and there the charter was formally accepted and the Company organized. As all the associates had not arrived till Saturday evening, the 6th of May, and as one of them was to leave town on Monday morning, it was agreed that they should meet for organization at six o'clock of that day. At that hour they came to the house of Mr. Field's brother Dudley, and as the first rays of the morning sun streamed into the windows, the formal organization took place. The charter was accepted, the stock subscribed, and the officers chosen. Mr. Cooper, Mr. Taylor, Mr. Field, Mr. Roberts, and Mr. White were the first directors. Mr. Cooper was chosen President, Mr. White, Vice-President, and Mr. Taylor, Treasurer.

This is a short story, and soon told. It seemed a

THE COMPANY ORGANIZED.

light affair, for half a dozen men to meet in the early morning and toss off such a business before breakfast. But what a work was that to which they thus put their hands! A capital of a million and a half of dollars was subscribed in those few minutes, and a company put in operation that was to carry a line of telegraph to St. John's, more than a thousand miles from New York, and then to span the wild sea. Well was it that they who undertook the work did not then fully realize its magnitude, or they would have shrunk from the attempt. Well was it for them that the veil was not lifted, which shut from their eyes the long delay, the immense toil, and the heavy burdens of many wearisome years. Such a prospect might have chilled the most sanguine spirit. But a kind Providence gives men strength for their day, imposes burdens as they are able to bear them, and thus leads them on to greater achievements than they knew.

CHAPTER IV.

CROSSING NEWFOUNDLAND.

There is nothing in the world easier than to build a line of railroad, or of telegraph, *on paper*. You have only to take the map, and mark the points to be connected, and then with a single sweep of the pencil to draw the line along which the iron track is to run. In this airy flight of the imagination, distances are nothing. A thousand leagues vanish at a stroke. All obstacles disappear. The valleys are exalted, and the hills are made low, soaring arches span the mountain streams, and the chasms are leaped in safety by the fire-drawn cars.

Very different is it to construct a line of railroad or of telegraph in reality; to come with an army of laborers, with axes on their shoulders to cut down the forests, and with spades in their hands to cast up the highway. Then poetry sinks to prose, and instead of flying over the space on wings, one must traverse it on foot, slowly and with painful steps. Nature asserts her power; and, as if resentful of the disdain with which man in his pride affected to leap over her, she piles up new barriers in his way. The mountains with their rugged sides cannot be moved out of their place, the rocks must be cleft in twain, to open a passage for the conqueror, before he can begin his triumphal march. The woods thicken into an impassable jungle; and the morass sinks deeper, threatening to swallow up the horse and his rider; until the rash projector is startled at his own audacity. Then it becomes a contest of forces between man and nature, in which, if he would be victorious, he must fight his way. The barriers of nature cannot be lightly pushed aside, but must yield at last only to time and toil, and " man's unconquerable will."

Seldom have all these obstacles been combined in a more formidable manner to obstruct any public work, than against the attempt to build a telegraph line across the island of Newfoundland. The distance, by the route to be traversed, was over four hundred miles, and the country was a wilderness, an utter desolation. Yet through such a country, over mountain and moor, through tangled brake and rocky gorge, over rivers and through morasses, they were to build a road—not merely a line of telegraph stuck on poles, but "a good and traversable bridle-road, eight feet wide, with bridges of the same width," from end to end of the island.

But nothing daunted, the new Company undertook the great work with spirit and resolution. Gisborne had made a beginning, and got some thirty or forty miles out of St. John's. This was the easiest part of the whole route, being in the most inhabited region of the island. But here he broke down, just where it was necessary to leave civilization behind, and to plunge into the wilderness.

Intending to resume the work on a much larger scale, Mr. White, the Vice-President, was sent down to St. John's to be the General Agent of the Company; while Mr. Matthew D. Field, as a practical engineer, was to have charge of the construction of the line. The latter soon organized a force of six hundred men, which he pushed forward in detachments to the scene of operations.

And now began to appear still more the difficulties of the way. To provide subsistence for man and beast, it was necessary to keep near the coast, for all supplies had to be sent round by sea. Yet in following the coast line, they had to wind around bays, or to climb over headlands. If they struck into the interior, they had to cut their way through the dense and tangled wood. There was not a path to guide them, not even an Indian trail. When lost in the forest, they had to follow the compass, as much as the mariner at sea.

To keep such a force in the field, that, like an army, produced nothing, but consumed fearfully, required constant attention to the commissary department. The little steamer Victoria, which belonged to the Company, was kept plying along the coast, carrying barrels of pork and potatoes, kegs of powder, pickaxes and spades and shovels, and all the implements of labor. These were taken up to the heads of the bays, and thence carried, chiefly on men's backs, over the hills to the line of the road.

In many respects, it had the features of a military expedition. It moved forward in a great camp. The men were sheltered in tents, when sheltered at all, or in small huts which they built along the road. But more often they slept on the ground. It was a wild and picturesque sight to come upon their camp in the woods, to see their fires blazing at night while hundreds of stalwart sleepers lay stretched on the ground. Sometimes, when encamped on the hills, they could be seen afar off at sea. It made a pretty picture then. But the hardy pioneers thought little of the figure they were making, when they were exposed to the fury of the elements. Often the rain fell in torrents, and the men, crouching under their slight shelter, listened sadly to the sighing of the wind among the trees, answered by the desolate moaning of the sea.

Yet in spite of all obstacles, the work went on. All through the long days of summer, and through the months of autumn, every cove and creek along that southern coast heard the plashing of their oars, and the steady stroke of their axes resounded through the forest.

But as the season advanced, all these difficulties increased. For nearly half the year, the island is buried in snow. Blinding drifts sweep over the moors, and choke up the paths of the forest. How at such times the expedition lay floundering in the woods, still struggling to force its way onward; what hardships and sufferings the men endured—all this is a chapter in the History of the Telegraph which has not been written, and which can never be fully told. The

> Gentlemen of England,
> Who dwell at home at ease,

and who are justly proud of the extent of their dominions, and the life and power which pervade the whole, may here find another example of the way in which great works are borne forward in distant parts of their empire.

But to carry out such an enterprise, requires head-work as well as hand-work. Engineering in the field must be supported by financiering at home. It was here the former enterprise broke down, and now it needed constant watching to keep the wheels in steady motion. The directors in New York found the demand increasing day by day. The minds which had grasped the large design must now descend to an infinity of detail. They had to keep an army of men at work, at a point a thousand miles away, far beyond their immediate oversight. Drafts for money came thick and fast. To provide for all these required constant attention. How faithfully they gave to this enterprise, not only their money, but their time and thought, few will know; but those who have seen can testify. In the autumn of that year, 1854, the writer removed to the city of New York, and was almost daily at the house of Mr. Field. Yet for months it was hardly possible to go there of an evening without finding the library occupied by the Company. Indeed, so uniformly was this the case, that " The Telegraph " began to be regarded by the family as an unwelcome intruder, since it put an interdict on the former social evenings and quiet domestic enjoyment. The circumstance shows the ceaseless care on the part of the directors which the enterprise involved. As a witness of their incessant labor, it is due to them to bear this testimony to their patience and their fidelity.

When they began the work, they hoped to carry the line across Newfoundland in one year, completing it in the summer of 1855. In anticipation of this, Mr. Field was sent by the Company to England at the close of 1854, to order a cable to span the Gulf of St. Lawrence, to connect Cape Ray with the island of Cape Breton. This was his first voyage across the ocean on the business of the Telegraph—to be followed by more than forty others. In London he met for the first time Mr. John W. Brett, with whom he was to be afterward connected in the larger enterprise of the

Atlantic Telegraph. Mr. Brett was the father of submarine telegraphy in Europe, though in carrying out his first projects he was largely indebted to Mr. Crampton, a well-known engineer of London, who aided him both with advice and capital. With this invaluable assistance, he had stretched two lines across the British channel. From his success in passing these waters, he believed a line might yet be stretched from continent to continent. The scientific men of England were not generally educated up to that point. The bare suggestion was received with a smile of incredulity.* But Mr. Brett had faith, even at that early day, and entered heartily into the schemes of Mr. Field. To show his interest, he afterward took a few shares in the Newfoundland line—the only Englishman who had any part in this preliminary work.

The summer came, and the work in Newfoundland, though not complete, was advancing; and the cable

---

* One or two exceptions there were, not to be forgotten. Professor William Thomson, of the University of Glasgow, then a young man, but full of the enthusiasm of science, was already prepared to welcome such a project, with confidence of success. As early as October and November, 1854, he wrote to the Secretary of the Royal Society of London, declaring his belief in its practicability. The letters are published in the Proceedings of the Royal Society for 1855. Such faith was not visionary, for it was based on clearer knowledge and more thorough investigation, and gave promise of those eminent services which this gentleman was afterwards to render to the cause of electrical science. Mr. C. F. Varley, also, was one of the first to perceive the possibility of an ocean telegraph, as he was to contribute greatly to its final success.

in England was finished and shipped on board the bark Sarah L. Bryant to cross the sea. Anticipating its arrival, the Company chartered a steamer to go down to Newfoundland to assist in its submersion across the Gulf of St. Lawrence. As yet they had no experience in the business of laying a submarine telegraph, and did not doubt that the work could be accomplished with the greatest ease. It was therefore to be an excursion of pleasure as well as of business, and accordingly they invited a large party to go with them to witness the unaccustomed spectacle.

As we chanced to be among the guests, we have the best reason to remember it. Seldom has a more pleasant party been gathered for any expedition. Representing the Company were Mr. Field, Mr. Peter Cooper, Mr. Robert W. Lowber, and Professor Morse; while among the invited guests were gentlemen of all professions—clergymen, doctors and lawyers, artists and editors. In the groups on the deck were the venerable Dr. Gardiner Spring and Rev. J. M. Sherwood; Dr. Lewis A. Sayre, Bayard Taylor, the well-known traveller, Mr. Fitz-James O'Brien, and Mr. John Mullaly—the three latter gentlemen representing leading papers of New York.* Besides these, the party included a

* The letters of Mr. Taylor, which first appeared in The New York Tribune have been since collected in one of his volumes of travel. Mr. O'Brien, a very brilliant writer, who afterward fell in our civil war, fighting bravely for his adopted country, furnished some spirited letters to The Times. But Mr. Mullaly, who appeared for The Herald, was the

large number of ladies, who gave life and animation to the company.

Well does the writer recall the morning of departure—the seventh day of August, 1855. Never did a voyage begin with fairer omens. It was a bright summer day. The sky was clear, and the water smooth. We were on the deck of the good ship James Adger, long known as one of the fine steamers belonging to the Charleston line. She was a swift ship, and cut the water like an arrow. Thus we sped down the bay, and turning into the ocean, skimmed along the shores of Long Island. The sea was tranquil as a lake. The whole party were on deck, scattered in groups here and there, watching the sails and the shore. A rude telegraph instrument furnished entertainment and instruction, especially as we had Professor Morse to explain his marvellous invention, which some who listened then for the first time understood.

At Halifax, several of us left the ship, and came across Nova Scotia, passing through that lovely region

most persevering attendant on the Telegraph, and the most indefatigable correspondent. He accompanied not only this expedition, but several others. He was on board the Niagara in 1857, and again in both the expeditions of 1858; and on the final success of the cable, prepared a volume, which was published by the Appletons, giving a history of the enterprise. This contains the fullest account of all those expeditions which has been given to the public. I have had frequent occasion to refer to his book, and can bear witness to the interest of the narrative. It is written with spirit, and doubtless would have had a longer life, if the cable itself had not come to an untimely end.

of Acadia which Longfellow has invested with such tender interest in his poem of Evangeline. Thence we crossed the Bay of Fundy to St. John in New Brunswick, and returned by way of Portland.

The James Adger went on to Newfoundland, steering first for Port au Basque, near Cape Ray, where they hoped to meet the bark which was to come from England with the cable on board. To their disappointment, it had not arrived. Mr. Canning, the engineer who was to lay the cable, had come out by steamer, and was on hand, but the bark was not to be seen. Having to wait several days, and wishing to make the most of their time, they sailed for St. John's, where they were received by the Provincial Government and the people with unbounded hospitality, after which they returned to Port au Basque, and were now rejoiced to discover the little bark hidden behind the rocks. It was decided to land the cable in Cape Ray Cove. After a day or two's delay in getting the end to the shore, they started to cross the Gulf of St. Lawrence, the Adger towing the bark. The sea was calm, and though they were obliged to move slowly, yet all promised well, till they were about half-way across, when a gale arose, which pitched the bark so violently, that with its unwieldy bulk it was in great danger of sinking. After holding on for hours in the vain hope that it would abate, the captain cut the cable to save the bark; and thus, after they had paid out forty miles, it was hopelessly lost, and the Adger returned to New York.

This loss was owing partly to the severity of the gale, and partly to the fact that the bark which had the cable on board was wholly unfitted for the purpose. It was a sailing-vessel, and had to be towed by another ship. In this way it was impossible to regulate its motion. It was too fast or too slow. It was liable to be swayed by the sea, now giving a lurch ahead, and now dragging behind. Experience showed that a cable should always be laid from a steam vessel which could regulate its own motion, running out freely when all went smoothly, and checking its speed instantly when it was necessary to ease up the strain, or to pay out more slack to fill up the hollows of the sea.

This first loss of a submarine cable was a severe disappointment to the Company. It postponed the enterprise for a whole year. To make a new cable would require several months, and the season was so far advanced that it could not be laid before another summer. Was it strange if some of the little band began to ask if they had not lost enough, and to reason that it was better to stop where they were, than to go on still farther, casting their treasures into the sea?

But there was in that little company a spirit of hope and determination that could not be subdued; that

ever cried: " Once more unto the breach, good friends!"
After some deliberation, it was resolved to renew the
attempt. Mr. Field again sailed for England to order
another cable, which was duly made and sent out the
following summer. This time, warned by experience,
the Company invited no party and made no display.
The cable was placed on board a steamer fitted for the
purpose; from which it was laid without accident, and
remained in perfect working order for nine years.

Meanwhile the work on land had been pushed forward without ceasing. After incredible labor, the
Company had built a road and a telegraph from one
end of Newfoundland to the other, four hundred miles;
and, as if that were not enough, had built also another line, one hundred and forty miles in length, in
the island of Cape Breton. The first part of their
work was now done. The telegraph had been carried
beyond the United States through the British Provinces to St. John's in Newfoundland, a distance from
New York of over one thousand miles.

The cost of the line, thus far, had been about a
million of dollars, and of this the whole burden, with
but trifling exceptions, had fallen upon the original
projectors—Mr. Field having put in over two hundred
thousand dollars in money—and Mr. Cooper, Mr.
Taylor, and Mr. Roberts each a little less. No other
contributors beyond the six original subscribers had
come, except Professor Morse, Mr. Robert W. Low

ber, Mr. Wilson G. Hunt, and Mr. John W. Brett. The list of directors and officers remained as it was at first, except that this year, 1856, Mr. White died, and his place as director was filled by Mr. Hunt, and that Mr. Field was chosen Vice-President, and Mr. Lowber Secretary. In all the operations of the Company thus far, the various negotiations, the plan of the work, the oversight of its execution, and the correspondence with the officers and others, mainly devolved upon Mr. Field.

And so at length, after two long and weary years, these bold projectors had accomplished half their work. They had passed over the land, and under the Gulf of St. Lawrence, and having reached the farthest point of the American coast, they now stood upon the cliffs of Newfoundland, looking off upon the wide sea.

CHAPTER V.

THE DEEP-SEA SOUNDINGS.

When a landsman, born far away among the mountains, comes down to the coast, and stands for the first time on the shore of the sea, it excites in him a feeling of awe and wonder, not unmingled with terror. There it lies, a level surface, with nothing that lifts up its head like a peak of his native hills. And yet it is so vast, stretching away to the horizon, and all over the sides of the round world; with its tides and currents that sweep from the equator to the pole; with its unknown depths and its ceaseless motion; that it is to him the highest emblem of majesty and of power—a not unworthy symbol of God himself.

In proportion to its mystery is the terror which hangs over it. A vague dread always surrounds the unknown. And what so unknown as the deep, unfathomable sea? For thousands of years the sails of ships, like winged birds, have skimmed over it, yet it has remained the one thing in nature beyond alike man's knowledge and his power:

> Man marks the earth with ruin,
> His control stops with the shore.

And the little that has been known of the ocean has been chiefly of its surface, of the winds that blow over it, and the waves that are lifted up on high. We knew somewhat of its tides and currents as observed in different parts of the earth. We saw off our coast the great Gulf Stream—that steady flow of waters, so mighty and mysterious, which, issuing out of the tropical regions, poured its warm current, sixty miles broad, right through the cold waters of the North Atlantic; and sweeping round, sent the airs of a softer climate over all the countries of Western Europe. Old voyagers told us of the trade-winds that blew across the Pacific, and of terrible monsoons in China and Indian seas. But all that did not reveal what was going on a hundred fathoms below the surface. These old sailors had marvellous tales of Indian pearl-divers, who, holding their breath, plunged to the depth of a few hundred feet; but they came up half-dead, with but little to tell except of the frightful monsters of the deep. The diving-bell was let down over sunken wrecks, but the divers came up only with tales of riches and ruin, of gold and gems and dead men's bones that lay mingled together on the deep sea floor. Was the bottom of the sea all like this? Was it a vast realm of death, the sepulchre of the world? No man could tell us. Poets might sing of the caves of ocean, but no eye of science had yet penetrated those awful depths, which the storms never reach.

THE DEEP-SEA SOUNDINGS.

It is indeed marvellous how little was known, up to a very recent date, of the true character of the ocean. Navigators had often tried to find out how deep it was. When lying becalmed on a tranquil sea, they had amused themselves by letting down a long line, weighted with a cannon-ball, to see if they could touch bottom. But the results were very uncertain. Sometimes the line ran out for miles and miles, but whether it was all the while descending, or was swayed hither and thither by mighty under-currents, could not be known.

But this true character of the ocean it was necessary to determine, before it could be possible to pass the gulf of the Atlantic. What was there on the bottom of the sea, where the cable was to find its resting place? Was that ocean-bed a wide level plain, or had it been heaved up by volcanic forces into a hundred mountain-peaks, with many a gorge and precipice between? Such was the character of a part of the basin of the ocean. Here and there, all over the globe, are islands, like the Peak of Teneriffe, thrown up in some fierce bursting of the crust of our planet, that shoot up in tremendous cliffs from the sea. Who shall say that the same cliffs do not shoot down below the waves a thousand fathoms deep? And might there not be such islands, which did not show their heads above the surface, lying in the track between Europe and America; or perchance a succession of mountain ranges, over which the cable would have to be stretched, and where hanging from the heights it would swing with the tide, till at last it snapped and fell into the abyss below? Such at least were possible dangers to be encountered; and it was not safe to advance a step till the basin of the North Atlantic was explored.

The progress of invention, so rapid on land, at length found a way of penetrating the sea, and even of turning up its bottom to the gaze of men. To measure the depth with something like mathematical accuracy, an instrument was introduced known among nautical men as Massey's Indicator, the method of which is very clearly explained in an article which appeared in one of the New York papers, (The Times,) on the deep-sea soundings made for the Atlantic Telegraph:

"The old system is with a small line, marked at distances of one hundred fathoms, and with a weight of thirty or fifty pounds, the depth being told by the length of line run out. This is, of course, the most natural apparatus that suggests itself, and has been in use from the earliest ages. Experience has given directions for its use, avoiding some of the grosser causes of error from driftage and other causes. Yet its success in immense ocean depths is problematical, and a problem decided in the negative by many of the first scientific authorities at home and abroad. In the mechanical improvements of the last half-century substitutes for this simple but rather uncertain method began to be devised. It

BROOKE'S DEEP SEA SOUNDING APPARATUS.

A shows the instrument ready for sounding. It is very simple, consisting only of a cannon-ball, pierced with an iron rod, and held in its place by slings. As the ball goes down swiftly, it drives the rod into the bottom like the point of a spear, when an opening at the end catches the ooze in its iron lips. The same instant (see B) the slings loosen, the ball drops off, and the naked rod, C, with its "bite" is drawn up to the surface.

was proposed to ascertain the depth by the amount of pressure, or by explosions under water, with other equally impracticable plans. At last was noticed the perfect regularity of the movements of a spirally-shaped wheel, on being drawn through the water. Experiments proved that this regularity, when unaffected by other causes, could be relied on with perfect accuracy, and that an arrangement of cog-wheels would register its revolutions with mathematical precision. Very soon it came in use as a ship's log. So perfect was their precision, that they were even introduced in scientific surveys. Base lines, where the nicest accuracy is required, were run with them, and we have the highest authority of the Royal Navy for believing that they never failed. At this point it was proposed to apply them in a perpendicular as well as in a horizontal motion through the water, Massey's apparatus promising to solve those problems of submarine geography left unsolved by the old method of obtaining depth with a simple line and sinker, and this more especially as some causes of error, considerable on the surface, disappear in the still water below."

To make our knowledge of the sea complete, one thing more was wanting—a method not only of reaching the bottom, but of laying hold of it, and bringing it up to the light of day. This was now to be supplied.

It is to the inventive genius of a lieutenant of the United States navy, Mr. J. M. Brooke, that the world owes the means of finding out what is at the bottom of the sea. This is by a very simple contrivance, by which the heavy weight, used to sink the measuring line, *is detached as soon as it strikes bottom,* leaving the line free so that it can be drawn up lightly and quickly to the surface without danger of breaking. Below the weight, and driven by it into the ooze, is a rod, in which is an open valve, that now closes with a spring, by which it catches a cupful of the soil, which is thus brought up to the surface, to be placed under the microscope, and be subjected to the sharp eye of science. With this simple instrument the skilful seaman explores the bottom of the ocean by literally feeling over it. With a long line he dives to the very lowest depths, while the clasp at the end of it is like the tip of the elephant's trunk, serving as a delicate finger with which he picks up sand and shells that lie strewn on the floor of the deep. What important conclusions are derived from this inspection of the bottom of the sea, is well stated by Lieutenant Maury in the letter already quoted.

In happy concurrence with this, as an additional preparation, a partial survey of the Atlantic had been made the very year before this enterprise was begun, in 1853. Lieutenant Berryman was the first who applied this new method of taking deep-sea soundings to that part of the Atlantic lying between Newfoundland and Ireland, with results most surprising and satisfactory. But to remove all doubt it seemed desirable to have a fresh survey. To obtain this, Mr. Field went to Washington and applied to the Government in behalf of the Company for a second expedition.

58   STORY OF THE ATLANTIC TELEGRAPH.

The request was granted, and the Arctic, under command of the same gallant Lieutenant Berryman, was assigned to this service. He sailed from New York on the eighteenth of July, 1856, and the very next day Mr. Field left on the Baltic for England, to organize the Atlantic Telegraph Company. The Arctic proceeded to St. John's, and thence with a clear eye and a steady hand, this true sailor went "sounding on his dim and perilous way" across the deep. In about three weeks he made the coast of Ireland, having carried his survey along the great circle arc, which the telegraph was to follow as the nearest path from the old world to the new. The result fully confirmed his belief of the existence of a great plateau underneath the ocean, extending all the way from one hemisphere to the other.

I cannot take leave of the name of this gallant officer, who rendered such services to science and to his country, without a word of tribute to his memory. Lieutenant Berryman is in his grave. He died in the navy of his country, worn out by his devotion to her service. When the great civil war broke out, he was placed in a position most painful to a man of large heart, who loved at once his country and the state in which he was born. He was a Southerner, a native of Winchester, Va., and was assigned to duty in the South. At the first attack on Southern forts and arsenals, he was in command of the Wyandotte, in the harbor of Pensacola, in Florida. His officers, who were nearly all Southerners, were in secret sympathy with the rebellion. All the influences around him, both on ship and on shore, were such as might have seduced a weaker man from his loyalty. But, to his honor, he never hesitated for a moment. He stood firm and loyal to his flag. Not knowing whom to trust, he had to keep watch day and night against surprise and treachery. It was the testimony of Lieutenant Slemmer, then in command of Fort Pickens, that but for the ceaseless exertions of Lieutenant Berryman not only the ship but the fort would have been lost. But this service to his country cost him his life. His constant exertions brought on a brain fever, of which he died. His wife, also a native of Winchester, when the war came near her early home, removed to Baltimore, saying that "she would not live under any other flag than that under which her husband had lived and died."

It was to the honor of the American navy; to have led the way in these deep-sea soundings. But after this second voyage of exploration, Mr. Field applied to the British Admiralty, "to make what further soundings might be necessary between Ireland and Newfoundland, and to verify those made by Lieutenant Berryman." It was in response to this application that the Government sent out the following year a vessel to make still another survey of the same ocean-path. This was the steamer Cyclops, which was placed under Lieutenant Commander Joseph Dayman, of the British navy, an officer who had been with Captain Sir James Ross when he made his deep-sea soundings in the South Atlantic in 1840, where he attained a depth of twenty-six hundred and sixty-seven fathoms; and who by his intelligence and zeal, was admirably fitted for the work. To speak now of this *third* survey, is anticipating in time. But it will serve the purpose of unity and clearness in the narrative, to include all these deep-sea soundings in one chapter. He was directed to proceed to the harbor of Valentia in Ireland, and thence to follow, as nearly as possible, along the arc of a great circle to Newfoundland. "The soundings for the first few miles from the coast should be frequent, decreasing as you draw off shore."

These orders were thoroughly executed. Every pains was taken to make the information obtained precise and exact. Whenever a sounding was to be taken, the ship was hove to, and the bow kept as nearly as possible in the same spot, so that the line might descend perpendicularly. This was repeated every few miles until they had got far out into the Atlantic, where the general equality of the depths rendered it necessary to cast the line only every twenty or thirty miles. Thus the survey was made complete, and the results obtained were of the greatest value in determining the physical geography of the sea.

THE DEEP-SEA SOUNDINGS. 61

The conclusions of Commander Dayman confirmed in general those of Lieutenant Berryman, though in comparing the charts prepared by the two, we observe some differences which ought to be noticed. Both agree as to the general character of the bottom of the ocean along this latitude—that it is a vast plain, like the steppes of Siberia. Yet on the chart of Dayman the floor of the sea seems *not such a dead level* as on that of Berryman. (This may be partly owing to a difference of route, as Dayman passed a little to the north of the track of Berryman.) There are more unequal depths, which in the small space of a chart, appear like hills and valleys. Yet when we consider the wide distances passed over, these inequalities seem not greater than the undulations on our Western prairies. "This space," says Dayman, "has been named by Maury the telegraphic plateau, and although by multiplying the soundings upon it, we have depths ranging from fourteen hundred and fifty to twenty-four hundred fathoms, these are comparatively small inequalities in its surface, and present no new difficulty to the project of laying the cable across the ocean. Their importance vanishes when the extent of the space over which they are distributed (thirty degrees of longitude) is considered."

According to Berryman and Dayman both, the ocean in its deepest part on this plateau, measured but two thousand and three or four hundred fathoms, or about

BED OF THE ATLANTIC, NORTH AND SOUTH, THROUGH THE CAPE DE VERDS, AZORES, AND TELEGRAPH PLATEAU.

fourteen thousand feet—a depth of but little over two and a half miles. This is not great, compared with the enormous depths in other parts of the Atlantic;* yet that it is *something* may be realized from the fact that if the Peak of Teneriffe were here "cast into the sea," it would sink out of sight, island, mountain and all, while even the lofty head of Mont Blanc would be lifted but a few hundred feet above the waves.

The only exception to this uniform depth, lies about two hundred miles off the coast of Ireland, where within a space of about a dozen miles, the depth sinks from five hundred and fifty to seventeen hundred and fifty fathoms! "In 14° 48' west," says Dayman, "we have five hundred and fifty fathoms rock, and in 15° 6' west we have seventeen hundred and fifty fathoms ooze. This is the greatest dip in the whole ocean."

* "The ocean bed of the North Atlantic is a curious study; in some parts furrowed by currents, in others presenting banks, the accumulations perhaps of the débris of these ocean rivers during countless ages. To the west, the Gulf Stream pours along in a bed from one mile to a mile and a half in depth. To the east of this, and south of the Great Banks, is a basin, eight or ten degrees square, where the bottom attains a greater depression than perhaps the highest peaks of the Andes or Himalayas—six miles of line have failed to reach the bottom! Taking a profile of the Atlantic basin in our own latitude, we find a far greater depression than any mountain elevation on our own continent. Four or five Alleghanies would have to be piled on each other, and on them added Fremont's Peak, before their point would show itself above the surface. Between the Azores and the mouth of the Tagus this decreases to about three miles."

"In little more than ten miles of distance a change of depth occurs, amounting to seventy-two hundred feet." This is indeed a tremendous plunge from the hard rock into the slime of the sea.

The same sharp declivity was noticed by Berryman, and has been observed in the several attempts to lay the cable. Thus in the second expedition of 1858, as the Agamemnon was approaching the coast of Ireland, we read in the report of her voyage: "About five o'clock in the evening, the steep submarine mountain which divides the telegraphic plateau from the Irish coast, was reached; and the sudden shallowing of the water had a very marked effect on the cable, causing the strain on, and the speed of it, to lessen every minute. A great deal of slack was paid out to allow for inequalities which might exist, though undiscovered by the sounding-line."

This submarine mountain was then regarded as the chief point of danger in the whole bed of the Atlantic, and as the principal source of anxiety in laying a cable across the ocean. Yet, after all, the ascent or descent of less than a mile and a half in ten miles, is not an impassable grade. More recent soundings reduce this still farther. Captain Hoskins, of the Royal Navy, afterwards made a more careful survey of this precipitous sea bottom, and with results much more favorable. The side of the mountain, it is now said, is not very much steeper than Holborn Hill in London, or Murray

THE DEEP-SEA SOUNDINGS. 65

Hill in New York.* But the best answer to fears on this point, is the fact that in 1857, 1858, and 1865, the cable passed over it without difficulty. In 1857 the Niagara was a hundred miles farther to sea, when the cable broke. In 1865 the strain was not increased more than a hundred pounds. In the final expedition, that of 1866, this declivity was passed over without difficulty or danger.

Next to the depth of the ocean, it was important to ascertain the nature of its bottom. What was it—a vast bed of rock, the iron-bound crust of the globe, hardened by internal fires, and which, bending as a

---

* The results obtained are thus summed up in the London Times:

"The dangerous part of this course has hitherto been supposed to be the sudden dip or bank which occurs off the west coast of Ireland, where the water was supposed to deepen in the course of a few miles from about three hundred fathoms to nearly two thousand. Such a rapid descent has naturally been regarded with alarm by telegraphic engineers, and this alarm has led to a most careful sounding survey of the whole supposed bank by Captain Dayman, acting under the instructions of the Admiralty. The result of this shows that the supposed precipitous bank, or submarine cliff, is a gradual slope of nearly sixty miles. Over this long slope the difference between its greatest height and greatest depth is only eighty-seven hundred and sixty feet; so that the average incline is, in round numbers, about one hundred and forty-five feet per mile. A good gradient on a railway is now generally considered to be one in one hundred feet, or about fifty-three in a mile; so that the incline on this supposed bank is only about three times that of an ordinary railway. In fact, as far as soundings can demonstrate any thing, there are few slopes in the bed of the Atlantic as steep as that of Holborn Hill. In no part is the bottom rocky, and with the exception of a few miles, which are shingly, only ooze, mud, or sand is to be found."

vault over the still glowing centre of the earth, bore up on its mighty arches the weight of all the oceans? or was it mere sand like the sea-shore? or ooze as soft as that of a mill-pond? The pressure of a column of water two miles high would be equal to that of four hundred atmospheres. Would not this weight alone be enough to crush any substance that could reach that tremendous depth? These were questions which remained to be answered, but on which depended the possibility of laying a cable at the bottom of the Atlantic.

By the ingenious contrivance of Lieutenant Brooke, the problem was solved, for we got hold of fragments of the under coating of the sea; and to our amazement, instead of finding the ocean bound round with thick ribs of granite, its inner lining was found to be soft as a silken vest. The soil brought up from the bottom was not even of the hardness of sand or gravel. It was mere ooze, like that of our rivers, and was as soft as the moss that clings to old, damp stones on the river's brink. At first it was thought by Lieutenant Berryman to be common clay, but being carefully preserved, and subjected to a powerful microscope, it was found to be composed of shells, too small to be discovered by the naked eye!

This was a revelation of the myriad forms of animated existence which fill the sea : a plenitude of life that is more wonderful by contrast. As Maury well puts it : " The ocean teems with life, we know. Of the four elements of the old philosophers—fire, earth, air, and water—perhaps the sea most of all abounds with living creatures. The space occupied on the surface of our planet by the different families of animals and their remains are inversely as the size of the individual. The smaller the animal, the greater the space occupied by his remains. Take the elephant and his remains, or a microscopic animal and his, and compare them. The contrast, as to space occupied, is as striking as that of the coral reef or island with the dimensions of the whale. The graveyard that would hold the corallines is larger than the graveyard that would hold the elephants."*

These little creatures, whose remains were thus found at the bottom of the ocean, probably did not live there, for there all is dark, and shells, like flowers, need the light and warmth of the all-reviving sun. It was their sepulchre, but not their dwelling-place. Probably they lived near the surface of the ocean, and after their short life, sunk to the tranquil waters below. What a work of life and death had been going on for ages in the depths of the sea! Myriads upon myriads, ever since the morning of creation, had been falling like snow-flakes, till their remains literally covered the bottom of the deep.

Equally significant was the fact that these shells

* Physical Geography of the Sea.

were *unbroken*. Not only were they there, but preserved in a perfect form. Organisms the most minute and delicate, fragile as drooping flowers, had yet sunk and slept uninjured. The same power which watches over the fall of a sparrow had kept these frail and tender things, and after their brief existence, had laid them gently on the bosom of the mighty mother for their eternal rest.

The bearing of this discovery on the problem of a submarine telegraph was obvious. For it too was to lie on the ocean-bed, beside and among these relics that had so long been drifting down upon the watery plain. And if these tiny shells slept there unharmed, surely an iron chord might rest there in safety. There were no swift currents down there; no rushing waves agitated that sunless sea. There the waters moved not; and there might rest the great nerve that was to pass from continent to continent. And so far as injury from the surrounding elements was concerned, there it might remain, whispering the thoughts of successive generations of men, till the sea should give up its dead.

CHAPTER VI.

THE WORK BEGUN IN ENGLAND.

Up to this time the Telegraph, which was destined to pass the sea, had been purely an American enterprise. It had been begun, and for over two years had been carried on, wholly by American capital. " Our little company," said Mr. Field ten years after, " raised and expended over a million and a quarter of dollars before an Englishman paid a single pound sterling." Mr. Brett was the first one to take a few shares. But this was not to the discredit of England, for the American public had done no better. Not a dollar had been raised this side the Atlantic, outside of the little circle in which the scheme had its origin. No stock or bonds were put upon the market; no man was asked for a subscription. If they wanted money, they drew their checks for it. At one time, indeed, two hundred and fifty thousand dollars of bonds were issued, but they were at once taken wholly by themselves. But, as the time was now come when the long-meditated attempt was to be made to carry the Telegraph across the ocean, it was fitting that Great Britain, whose shores it was to touch, should join in the work. Accordingly, in the summer of 1856, after finishing all that he could do in America, Mr. Field sailed with his family for England. The very day before he embarked, he had the pleasure to see his friend, Lieutenant Berryman, off on his second voyage to make soundings across the Atlantic.

In London he sought at once Mr. Brett, with whom in his two former visits to England he had already discussed his project, and found in him a hearty coöperator. As we go on with our story, it is a melancholy satisfaction to refer to one and another worker in this enterprise, who lived not to see its last and greatest triumph. Mr. Brett, like Berryman, is dead. But he did not go to his grave till after a life of usefulness and honor. He was one of the men of the new era—of the school of Stephenson and Brunel—who believed in the marvellous achievements yet to be wrought by human invention, turning to the service of man the wonders of scientific discovery. He was one of the first to see the boundless possibilities of the telegraph, and to believe that what had passed over the land might pass under the sea. He was the first to lay a cable across the British Channel, and thus to bring into instantaneous communication the two great capitals of Europe—an achievement which, though small compared with what has since been done, was then so marvellous, that the intelligence of its success

THE WORK BEGUN IN ENGLAND. 71

was received with surprise and incredulity. Many could not and would not believe it. Even after messages were received in London from Paris, there were those who declared that it was an imposition on the public, with as much proud scorn as some a few years later scouted the very idea that a message had ever passed over the Atlantic Telegraph!

This friendship of Mr. Brett—both to the enterprise and to Mr. Field personally—remained to the last. In every voyage to England the latter found—however others doubted or despaired—that Mr. Brett was always the same—full of hope and confidence. In 1864, when they met in London, he was unshaken in faith, and urgent to have the great enterprise renewed. The triumph was not far off, but he was not to live to see it. But, though he passed away before the final victory, he did his part toward bringing it on, and no history of this great enterprise can overlook his eminent services.

To Mr. Brett, therefore, Mr. Field went first to consult in regard to his project of a telegraph across the ocean. This was a part of the design embraced in the original organization of the New York, Newfoundland, and London Telegraph Company; and when Mr. Field went to England, he was empowered to receive subscriptions to that Company, so as to enlarge its capital, and thus include in one corporation the whole line from New York to London; or to organize a new company, which should lay a cable across the Atlantic, and there join the Newfoundland line.

But before an enterprise so vast and so new could be commended to the commercial public of Great Britain, there were many details to be settled. The mechanical and scientific problems already referred to, whether a cable could be laid across the ocean; and if so, whether it could be worked, were to be considered anew. The opinions of Lieutenant Maury and of Professor Morse were published in England, and arrested the attention of scientific men. But John Bull is slow of belief, and asked for more evidence. The thing was too vast to be undertaken rashly. As yet there was no experience to decide the possibility of a telegraph across the ocean. The longest line which had been laid was three hundred miles. This caution, which is a national trait of Englishmen, will not be regarded as a fault by those who consider that in proportion as they are slow to embark in any new enterprise, are they resolute and determined in carrying it out.

To resolve these difficult problems, Mr. Field sought counsel of the highest engineering authorities of Great Britain, and of her most eminent scientific men. To their honor, all showed the deepest interest in the project, and gave it freely the benefit of their knowledge.

First, as to the possibility of laying a cable in the deep sea, Mr. Field had witnessed one attempt of the kind—that in the Gulf of St. Lawrence the year before —an attempt which had failed. His experience, therefore, was not encouraging. If they found so much difficulty in laying a cable seventy miles long, how could they hope to lay one of two thousand miles across the stormy Atlantic?

This was a question for the engineers. To solve the problem, required experiments almost without number. It was now that the most important services were rendered by Glass, Elliot & Co., of London, a firm which had begun within a few years the manufacture of sea-cables, and was to write its name in all the waters of the world. Aided by the skill of their admirable engineer, Mr. Canning, they now manufactured cables almost without end, applying to them every possible test. At the same time, Mr. Field took counsel of Robert Stephenson and George Parker Bidder, both of whom manifested a deep interest in the success of the enterprise.

Not less cordial was Mr. Brunel, who made many suggestions in regard to the form of the cable, and the manner in which it should be laid. He was then building the Great Eastern; and one day he took Mr. Field down to Blackwall to see it, and, pointing to the monstrous hull which was rising on the banks of the Thames, said: "There is the ship to lay the Atlantic cable!" Little did he think that ten years after, that ship would be employed in this service; and in this final victory over the sea, would redeem all the misfortunes of her earlier career.

Among the difficulties to be encountered, was that of finding a perfect insulator. Without insulation, telegraphic communication by electricity is impossible. On land, where wires are carried on the tops of poles, the air itself is a sufficient insulator. A few glass rings at the points where the wire passes through the iron staples by which it is supported, and the insulation is complete. But in the sea the electricity would be instantly dissipated, unless some material could be found which should insulate a conductor sunk in water, as completely as if it were raised in air. But what could thus inclose the lightning, and keep it fast while flying from one continent to the other?

Here again it seemed as if Divine wisdom had anticipated the coming of this great enterprise, and provided in the realm of nature every material needed for its success. It was at least a happy coincidence that only a few years before there had been found, in the forests of the Malayan archipelago, a substance till then unknown to the world, but which answered completely this new demand. This was gutta-percha, which is impenetrable by water, and at the same time a bad conductor of electricity; so that it forms at once a perfect protection and insulation to a telegraph passing through the sea. In the experiments that were made to test the value of this material in the grander use to

THE WORK BEGUN IN ENGLAND. 75

which it was to be applied, no man rendered greater service than Mr. Samuel Statham, of the London Gutta-Percha Works—a name to be gratefully remembered in the early history of the Atlantic Telegraph.

The mechanical difficulties removed, and the insulation provided, there remained yet the great scientific problem: Could a message be sent two thousand miles under the Atlantic? The ingenuity of man might devise some method of laying a cable across the sea, but of what use were it, if the electric current should shrink from the dark abyss?

It was in prosecuting inquiries to resolve this problem, that Mr. Field became acquainted with two gentlemen who were to be soon after associated with him in the organization of the Atlantic Telegraph Company. These were Mr. Charles T. Bright, afterward knighted for his part in laying the Atlantic cable in 1858, and Dr. Edward O. Whitehouse, both well known in England, the former as an engineer, and the latter for his experiments in electro-magnetism, as applied to the business of telegraphing. He had invented an instrument by which to ascertain and register the velocity of electric currents through submarine cables. Both these gentlemen were full of the ardor of science, and entered on this new project with the zeal which the prospect of so great a triumph might inspire. With them was now to be associated our distinguished countryman, Professor Morse. Fortunately he was at this time in London, and gave his invaluable aid to the experiments which were made to determine the possibility of telegraphic communication at great distances under the sea. The result of his experiments he communicates in a letter to Mr. Field:

"LONDON, FIVE O'CLOCK A. M.,
" October 3, 1856.

"MY DEAR SIR : As the electrician of the New York, Newfoundland, and London Telegraph Company, it is with the highest gratification that I have to apprise you of the result of our experiments of this morning upon a single continuous conductor of more than two thousand miles in extent, a distance you will perceive sufficient to cross the Atlantic Ocean, from Newfoundland to Ireland.

"The admirable arrangements made at the Magnetic Telegraph Office in Old Broad street, for connecting ten subterranean gutta-percha insulated conductors, of over two hundred miles each, so as to give one continuous length of more than two thousand miles during the hours of the night, when the telegraph is not commercially employed, furnished us the means of conclusively settling, by actual experiment, the question of the practicability as well as the practicality* of telegraphing through our proposed Atlantic cable.

"This result had been thrown into some doubt by the dis-

---

* Professor Morse was fond of the distinction between the words practical and practicable. A thing might be practicable, that is, possible of accomplishment, when it was not a practical enterprise, that is, one which could be worked to advantage. He here argues that the Atlantic Telegraph is both practicable, (or possible,) and at the same time a wise, practical undertaking.

THE WORK BEGUN IN ENGLAND. 77

covery, more than two years since, of certain phenomena upon subterranean and submarine conductors, and had attracted the attention of electricians, particularly of that most eminent philosopher, Professor Faraday, and that clear-sighted investigator of electrical phenomena, Dr. Whitehouse ; and one of these phenomena, to wit, the perceptible retardation of the electric current, threatened to perplex our operations, and required careful investigation before we could pronounce with certainty the commercial practicability of the Ocean Telegraph.

"I am most happy to inform you that, as a crowning result of a long series of experimental investigation and inductive reasoning upon this subject, the experiments under the direction of Dr. Whitehouse and Mr. Bright, which I witnessed this morning—in which the induction coils and receiving magnets, as modified by these gentlemen, were made to actuate one of my recording instruments—have most satisfactorily resolved all doubts of the practicability as well as practicality of operating the telegraph from Newfoundland to Ireland.

"Although we telegraphed signals at the rate of two hundred and ten, two hundred and forty-one, and, according to the count at one time, even of two hundred and seventy per minute upon my telegraphic register, (which speed, you will perceive, is at a rate commercially advantageous,) these results were accomplished notwithstanding many disadvantages in our arrangements of a temporary and local character—disadvantages which will not occur in the use of our submarine cable.

"Having passed the whole night with my active and agreeable collaborators, Dr. Whitehouse and Mr. Bright, without sleep, you will excuse the hurried and brief character of this note, which I could not refrain from sending you, since our experiments this morning settle the scientific and commercial points of our enterprise satisfactorily.

"With respect and esteem, your obedient servant,

"SAMUEL F. B. MORSE."

A week later, he wrote again, confirming his former impressions, thus:

"LONDON, October 10, 1856.

"MY DEAR SIR: After having given the deepest consideration to the subject of our successful experiments the other night, when we signalled clearly and rapidly through an unbroken circuit of subterranean conducting wire, over two thousand miles in length, I sit down to give you the result of my reflections and calculations.

"There can be no question but that, with a cable containing a single conducting wire, of a size not exceeding that through which we worked, and with equal insulation, it would be easy to telegraph from Ireland to Newfoundland at a speed of at least from eight to ten words per minute; nay, more: the varying rates of speed at which we worked, depending as they did upon differences in the arrangement of the apparatus employed, do of themselves prove that even a higher rate than this is attainable. Take it, however, at ten words in the minute, and allowing ten words for name and address, we can safely calculate upon the transmission of a twenty-word message in three minutes;

"Twenty such messages in the hour;

"Four hundred and eighty in the twenty-four hours, or fourteen thousand four hundred words per day.

THE WORK BEGUN IN ENGLAND.

"Such are the capabilities of a single wire cable fairly and moderately computed.

"It is, however, evident to me, that by improvements in the arrangement of the signals themselves, aided by the adoption of a code or system constructed upon the principles of the best nautical code, as suggested by Dr. Whitehouse, we may at least double the speed in the transmission of our messages.

"As to the structure of the cable itself, the last specimen which I examined with you seemed to combine so admirably the necessary qualities of strength, flexibility, and lightness, with perfect insulation, that I can no longer have any misgivings about the ease and safety with which it will be submerged.

"In one word, the doubts are resolved, the difficulties overcome, success is within our reach, and the great feat of the century must shortly be accomplished.

"I would urge you, if the manufacture can be completed within the time, (and all things are possible now,) to press forward the good work, and not to lose the chance of laying it during the ensuing summer.

"Before the close of the present month, I hope to be again landed safely on the other side of the water, and I full well know, that on all hands the inquiries of most interest with which I shall be met, will be about the Ocean Telegraph.

"Much as I have enjoyed my European trip this year, it would have enhanced the gratification which I have derived from it more than I can describe to you, if on my return to America, I could be the first bearer to my friends of the welcome intelligence that the great work had been begun, by the commencement of the manufacture of the cable to connect Ireland with the line of the New York,

Newfoundland, and London Telegraph Company, now so successfully completed to St. John's.

"Respectfully, your obedient servant,

"SAMUEL F. B. MORSE."

These experiments and others removed the doubts of scientific men. Professor Faraday, in spite of the law of the retardation of electricity on long circuits, which it was said he had discovered, and which would render it impossible to work a line of such length as from Ireland to Newfoundland, now declared his full conviction that it was within the bounds of possibility. The passage of electricity might not be absolutely instantaneous, or have the swiftness of the solar beam, yet it would be rapid enough for all practical purposes. When Mr. Field asked him how long it would take for the electricity to pass from London to New York, he answered: "Possibly one second!"

Thus fortified by the highest scientific and engineering authorities, the projectors of an ocean telegraph were now ready to bring it before the British public, and to see what support could be found from the English Government and the English people.

Mr. Field first addressed himself to the Government. Without waiting for the Company to be fully organized, with true American eagerness and impatience, he wrote a letter to the Admiralty asking for a fresh survey of the route to be traversed, and for the aid of Government ships to lay the cable. He also addressed a letter to Lord Clarendon, stating the large design which they had conceived, and asking for it the aid which was due to what concerned the honor and interest of England. The reply was prompt and courteous, inviting him to an interview for the purpose of a fuller explanation. Accordingly, Mr. Field, with Professor Morse, called upon him at the Foreign Office, and spent an hour in conversation on the proposed undertaking. Lord Clarendon showed great interest, and made many inquiries. He was a little startled at the magnitude of the scheme, and the confident tone of the projectors, and asked pleasantly: "But, suppose you *don't* succeed? Suppose you make the attempt and fail—your cable is lost in the sea—then what will you do?" "Charge it to profit and loss, and go to work to lay another," was the quick answer of Mr. Field, which amused him as a truly American reply. In conclusion, he desired him to put his request in writing, and, without committing the Government, encouraged him to hope that Britain would do all that might justly be expected in aid of this great international work. How nobly this promise was kept, time will show.

While engaged in these negotiations, Mr. Field took his family to Paris, and there met with a great loss in the sudden death of a favorite sister, who had accompanied them abroad. Full of the sorrow of this event, and unfitted for business of any kind, he returned to

London to find an invitation to go into the country and spend a few days with Mr. James Wilson, then Secretary to the Treasury, a man of great influence in the Government, at his residence near Bath; there to discuss quietly and at length the proposed aid to the Atlantic Telegraph. Though he had but little spirit to go among strangers, he felt it his duty not to miss an opportunity to advance the cause he had so much at heart. The result of this visit was the following letter, received a few days later:

"TREASURY CHAMBERS, Nov. 20, 1856.

"SIR: Having laid before the Lords Commissioners of her Majesty's Treasury your letter of the 13th ultimo, addressed to the Earl of Clarendon, requesting, on behalf of the New York, Newfoundland, and London Telegraph Company, certain privileges and protection in regard to the line of telegraph which it is proposed to establish between Newfoundland and Ireland, I am directed by their lordships to acquaint you that they are prepared to enter into a contract with the said Telegraph Company, based upon the following conditions, namely:

"1. It is understood that the capital required to lay down the line will be three hundred and fifty thousand pounds.

"2. Her Majesty's Government engage to furnish the aid of ships to take what soundings may still be considered needful, or to verify those already taken, and favorably to consider any request that may be made to furnish aid by their vessels in laying down the cable.

"3. The British Government, from the time of the completion of the line, and so long as it shall continue in working order, undertakes to pay at the rate of fourteen thousand pounds a year, being at the rate of four per cent. on the assumed capital, as a fixed remuneration for the work done on behalf of the Government, in the conveyance outward and homeward of their messages. This payment to continue until the net profits of the Company are equal to a dividend of six per cent., when the payment shall be reduced to ten thousand pounds a year, for a period of twenty-five years.

"It is, however, understood that if the Government messages in any year shall, at the usual tariff-rate charged to the public, amount to a larger sum, such additional payment shall be made as is equivalent thereto.

"4. That the British Government shall have a priority in the conveyance of their messages over all others, subject to the exception only of the Government of the United States, in the event of their entering into an arrangement with the Telegraph Company similar in principle to that of the British Government, in which case the messages of the two Governments shall have priority in the order in which they arrive at the stations.

"5. That the tariff of charges shall be fixed with the consent of the Treasury, and shall not be increased, without such consent being obtained, as long as this contract lasts.

"I am, sir, your obedient servant,

"JAMES WILSON.

"CYRUS W. FIELD, Esq., 37 Jermyn street."

With this encouragement and promise of aid, the projectors of a telegraph across the ocean now went forward to organize a company to carry out their design. Mr. Field, on arriving in England, had entered into an agreement with Mr. Brett to join their efforts for this purpose. With them were afterward united two others—Sir Charles Bright, as engineer, and Dr. Whitehouse, as electrician. These four gentlemen agreed to form a new company, to be called The Atlantic Telegraph Company, the object of which should be "to continue the existing line of the New York, Newfoundland, and London Telegraph Company to Ireland, by making or causing to be made a submarine telegraph cable for the Atlantic."

As they were now ready to introduce the enterprise to the British public, Mr. Field issued a circular in the name of the Newfoundland Company, and as its Vice-President, setting forth the great importance of telegraphic communication between the two hemispheres.

The next step was to raise the capital. After the most careful estimates, it was thought that a cable could be made and laid across the Atlantic for £350,000. This was a large sum to ask from a public slow to move, and that lends a dull ear to all new schemes. But armed with facts and figures, with maps and estimates, with the opinions of engineers and scientific men, they went to work, not only in London, but in other parts of the kingdom. Mr. Field, in company with Mr. Brett, made a visit to Liverpool and Manchester, to address their Chambers of Commerce. I have now before me the papers of those cities, with reports of the meetings held and the speeches made, which show the vigor with which they pushed their enterprise. This energy was rewarded with success. The result justified their confidence. In a few weeks the whole capital was subscribed. It had been divided into three hundred and fifty shares of a thousand pounds each. Of these, a hundred and one were taken in London, eighty-six in Liverpool, thirty-seven in Glasgow, twenty-eight in Manchester, and a few in other parts of England. The grandeur of the design attracted public attention, and some subscribed solely from a noble wish to take part in such a work. Among these were Mr. Thackeray and Lady Byron. Mr. Field subscribed £100,000, and Mr. Brett £25,000. But when the books were closed, it was found that they had more money subscribed than they required, so that in the final division of shares, there were allotted to Mr. Field eighty-eight, and to Mr. Brett twelve. Mr. Field's interest was thus one-fourth of the whole capital of the Company.

In taking so large a share, it was not his intention to carry this heavy load alone. It was too large a proportion for one man. But he took it for his countrymen. He thought one fourth of the stock should be held in this country, and did not doubt, from the eagerness with which three fourths had been taken in England, that the remainder would be at once subscribed in America. Had he been able, on his return, to attend to his own interests in the matter, this expectation might have been realized; but, as we shall see, hardly did he set foot in New York, before he was obliged to hurry off to Newfoundland on the business of the Company, and when he returned the interest had subsided, so that it required very great exertions, continued through many months, to dispose of twenty-seven shares. Thus he was by far the largest stockholder in England or America—his interest being over seven times that of Mr. Brett, who was the largest next to himself—and being more than double the amount held by all the other American shareholders put together. This was at least giving substantial proof of his own faith in the undertaking.

But some may imagine that after all this burden was not so great as it seemed. In many stock companies the custom obtains of assigning to the projectors a certain portion of the stock as a bonus for getting up the company, which amount appears among the subscriptions to swell the capital. It is indeed subscribed, *but not paid*. So some have asked whether this large subscription of Mr. Field was not in part at least merely nominal? To this we answer, that a consideration *was* granted to Mr. Field and his associates for their services in getting up the Company, and for their exclusive rights, but this was a contingent interest in the profits of the enterprise, *to be allowed only after the cable was laid*. So that the whole amount here subscribed was a *bona-fide* subscription, and paid in solid

English gold. We have now before us the receipts of the bankers of the Company for the whole amount, eighty-eight thousand pounds sterling.

The capital being thus raised, it only remained to complete the organization of the Company by the choice of a Board of Directors, and to make a contract for the cable. The Company was organized in December, 1856, by the choice of Directors chiefly from the leading bankers and merchants of London and Liverpool. The list included such honored names as Samuel Gurney, T. H. Brooking, John W. Brett, and T. A. Hankey, of London; Sir William Brown, Henry Harrison, Edward Johnston, Robert Crosbie, George Maxwell, and C. W. H. Pickering, of Liverpool; John Pender and James Dugdale, of Manchester; and Professor William Thomson, LL.D., of Glasgow. With these English Directors were two of our countrymen, Mr. George Peabody and Mr. C. M. Lampson, who, residing abroad for more than a third of a century, did much in the commercial capital of the world to support the honor of the American name. Mr. Peabody's firm subscribed £10,000, and Mr. Lampson £2,000. The latter gave more time than any other Director in London, except Mr. Brooking, the second Vice-Chairman, who, however, retired from the Company after the first failure in 1858, when Mr. Lampson was chosen to fill his place. The whole Board was full of zeal and energy. All gave their services without compensation.

It was the good fortune of the Company to have, from the beginning, in the important position of Secretary, a gentleman admirably qualified for the post. This was Mr. George Saward—a name familiar to all who have followed the fortunes of the telegraph, in England or America, since he has been the organ of communication with the press and the public; and with whom none ever had occasion to transact business without recognizing his intelligence and courtesy.

The Company being thus in working order, proceeded to make a contract for the manufacture of a cable to be laid across the Atlantic. For many months the proper form and size of the cable had been the subject of constant experiments. The conditions were: to combine the greatest degree of strength with lightness and flexibility. It must be strong, or it would snap in the process of laying. Yet it would not do to have it too large, for it would be unmanageable. Mr. Brett had already lost a cable in the Mediterranean chiefly from its bulk. Its size and stiffness made it hard to unwind it, while its enormous weight, when once it broke loose, caused it to run out with fearful velocity, till it was soon lost in the sea. It was only the year before, in September, 1855, that this accident had occurred in laying the cable from Sardinia to Algeria. All was going on well, until suddenly, "about two miles, weighing sixteen tons, flew out with the greatest violence in four or five minutes, flying round

THE WORK BEGUN IN ENGLAND. 89

even when the drums were brought to a dead stop, creating the greatest alarm for the safety of the men in the hold and for the vessel." This was partly owing to the character of the submarine surface over which they were passing. The bottom of the Mediterranean is volcanic, and is broken up into mountains and valleys. The cable, doubtless, had just passed over some Alpine height, and was descending into some fearful depth below; but chiefly it was owing to the great size and bulk of the cable. This was a warning to the Atlantic Company. The point to be aimed at was to combine the flexibility of a common ship's rope with the tenacity of iron. These conditions were thought to be united in the form that was adopted.* A con-

* On his return to America, many inquiries were addressed to Mr. Field in regard to the form and structure of the cable, in answer to which he wrote a letter of explanation in which he said :

"No particular connected with this great project has been the subject of so much comment through the press as the form and structure of the telegraph cable. It may be well believed that the Directors have not decided upon a matter so all-important to success, without availing themselves of the most eminent talent and experience which could be commanded. The practical history of submarine telegraphs dates from the successful submersion of the cable between Dover and Calais in 1851, and advantage has been taken of whatever instruction this history could furnish or suggest. Of the submarine cables heretofore laid down, without enumerating others, the one between Dover and Calais weighs six tons to the mile; that between Spezzia and Corsica, eight tons to the mile; that laid from Varna to Balaklava, and used during the war in the Crimea, less than three hundred pounds to the mile; while the weight of the cable for the Atlantic Telegraph is between nineteen hundred pounds and one ton to the mile. This cable, to use the words of Dr.

tract was at once made for the manufacture of the cable, one half being given to Messrs. Glass, Elliot & Co., of London, and the other to Messr. R. S. Newall & Co., of Liverpool. The whole was to be completed by the first of June, ready to be submerged in the sea. The company was organized on the ninth of December, and the very next day Mr. Field sailed for America, reaching New York on the twenty-fifth of December, after an absence of more than five months.

Whitehouse, 'is the result of many months thought, experiment, and trial. Hundreds of specimens have been made, comprising every variety of form, size, and structure, and most severely tested as to their powers and capabilities; and the result has been the adoption of this, which we know to possess all the properties required, and in a far higher degree than any cable that has yet been laid. Its flexibility is such as to make it as manageable as a small line, and its strength such that it will bear, in water, over six miles of its own weight suspended vertically.' The conducting medium consists not of one single straight copper-wire, but of seven wires of copper of the best quality, twisted round each other spirally, and capable of undergoing great tension without injury. This conductor is then enveloped in three separate coverings of gutta-percha, of the best quality, forming the core of the cable, round which tarred hemp is wrapped, and over this, the outside covering, consisting of eighteen strands of the best quality of iron-wire; each strand composed of seven distinct wires, twisted spirally, in the most approved manner, by machinery specially adapted to the purpose. The attempt to insulate more than one conducting-wire or medium would not only have increased the chances of failure of all of them, but would have necessitated the adoption of a proportionably heavier and more cumbrous cable. The tensile power of the outer or wire covering of the cable, being very much less than that of the conductor within it, the latter is protected from any such strain as can possibly rupture it or endanger its insulation without an entire fracture of the cable."

CHAPTER VII.

SEEKING AID FROM CONGRESS.

WHEN Mr. Field reached home from abroad, he hoped for a brief respite. He had had a pretty hard campaign during the summer and autumn in England, and needed at least a few weeks of rest; but that was denied him. He landed in New York on Christmas Day, and was not allowed even to spend the New Year with his family. There were interests of the Company in Newfoundland which required immediate attention, and it was important that one of the Directors should go there without delay. As usual, it devolved upon him. He left at once for Boston, where he took the steamer to Halifax, and thence to St. John's. Such a voyage may be very agreeable in summer, but in mid-winter it is not a pleasant thing to face the storms of those northern latitudes. The passage was unusually tempestuous. At St. John's he broke down, and was put under the care of a physician. But he did not stop to think of himself. The work for which he came was done; and though the physician warned him that it was a great risk to leave his bed, he took the steamer on her return, and was again in

New York after a month's absence—a month of hardship, of exposure, and of suffering, such as he had long occasion to remember.

The mention of this voyage came up a year afterward at a meeting of the Atlantic Telegraph Company in London, when a resolution was offered, tendering Mr. Field a vote of thanks for "the great services he had rendered to the Company by his untiring zeal, energy, and devotion." Mr. Brooking, the Vice-Chairman, had spent a large part of his life in Newfoundland, and knew the dangers of that inhospitable coast, and in seconding the resolution he said:

"It is now about a year and a half ago since I had the pleasure of making the acquaintance of my friend Mr. Field. It was he who initiated me into this Company, and induced me to take an interest in it from its earliest stage. From that period to the present I have observed in Mr. Field the most determined perseverance, and the exercise of great talent, extraordinary assiduity and diligence, coupled with an amount of fortitude which has seldom been equalled. I have known him cross the Atlantic in the depth of winter, and, within twenty-four hours after his arrival in New York, having ascertained that his presence was necessary in a distant British colony, he has not hesitated at once to direct his course thitherward. That colony is one with which I am intimately acquainted, having resided in it for upward of twenty years, and am enabled to speak of the hazards and danger which attend a voyage to it in winter. Mr. Field no sooner arrived at New York, in the latter part of December, than he got aboard a steamer for

Halifax, and proceeded to St. John's, Newfoundland. In three weeks he accomplished there a very great object for this Company. He procured the passage of an Act of the Legislature which has given to our Company the right of establishing a footing on those shores. [The rights before conferred, it would seem, applied only to the Newfoundland Company.] That is only one of the acts which he has performed with a desire to promote the interests of this great enterprise."

The very next day after his return from Newfoundland, Mr. Field was called to Washington, to seek the aid of his own Government to the Atlantic Telegraph. The English Government had proffered the most generous aid, both in ships to lay the cable, and in an annual subsidy of £14,000. It was on every account desirable that this should be met by corresponding liberality on the part of the American Government. Before he left England, he had sent home the letter received from the Lords Commissioners of the Treasury; and thereupon the Directors of the New York, Newfoundland, and London Telegraph Company had inclosed a copy to the President, with a letter asking for the same aid in ships, and in an annual sum of $70,000, [equivalent to £14,000,] to be paid for the government messages, the latter to be conditioned on the success of the telegraph, and to be continued only so long as it was in full operation. They urged with reason that the English Government had acted with great liberality—not only toward the enterprise, but toward our own Government. Although both ends of the line were in the British possessions, it had claimed no exclusive privileges, but had stipulated for perfect equality between the United States and Great Britain. The agreement expressly provided " that the British Government shall have a priority in the conveyance of their messages over all others, *subject to the exception only of the Government of the United States*, in the event of their entering into an arrangement with the Telegraph Company similar in principle to that of the British Government, in which case the messages of the two governments shall have priority in the order in which they arrive at the stations."

The letter to the President called attention to this generous offer—an offer which it was manifestly to the advantage of our Government to accept—and added: " The Company will enter into a contract with the Government of the United States on the same terms and conditions as it has made with the British Government." They asked only for the same recognition and aid which they had received in England. This surely was not a very bold request. It was natural that American citizens should think that in a work begun by Americans, and of which, if successful, their country would reap largely the honor and the advantage, they might expect the aid from their own Government which they had already received from a foreign power. It was, therefore, not without a mixture of surprise and mortification that they learned that the proposal in Congress had provoked a violent opposition, and that the bill was likely to be defeated. Such was the attitude of affairs when Mr. Field returned from Newfoundland, and which led him to hasten to Washington.

He now found that it was much easier to deal with the English than with the American Government. Whatever may be said of the respective methods of administration, it must be confessed that the forms of English procedure furnish greater facility in the despatch of business. A contract can be made by the Lords of the Treasury without waiting the action of Parliament. The proposal is referred to two or three intelligent officers of the Government—perhaps even to a single individual—on whose report it takes action without further delay. Thus it is probable that the action of the British Government was decided wholly by the recommendation of Mr. Wilson, formed after the visit of Mr. Field.

But in our country we do things differently. Here it would be considered a stretch of power for any administration to enter into a contract with a private company—a contract binding the Government for a period of twenty-five years, and involving an annual appropriation of money—without the action of Congress. This is a safeguard against reckless and extravagant expenditure, but, as one of the penalties we pay for our more popular form of government, in which every thing has to be referred to the people, it involves delay, and sometimes the defeat of wise and important public measures.

Besides—shall we confess it to our shame—another secret influence often appears in American legislation, which has defeated many an act demanded by the public good—the influence of the Lobby! This now began to show itself in opposition. It had been whispered in Washington that the gentlemen in New York who were at the head of this enterprise were very rich; and a measure coming from such a source surely ought to be made to pay tribute before it was allowed to pass. This was a new experience. Those few weeks in Washington were worse than being among the icebergs off the coast of Newfoundland. The Atlantic Cable has had many a kink since, but never did it seem to be entangled in such a hopeless twist as when it got among the politicians.

But it would be very unjust to suppose that there were no better influences in our Halls of Congress. There were then—as there have always been in our history—some men of large wisdom and of a noble patriotic pride, who in such a measure thought only of the good of their country and of the triumph of science and of civilization.

Two years after—in August, 1858—when the Atlantic Telegraph proved at last a reality, and the

New World was full of its fame, Mr. Seward, in a speech at Auburn, thus referred to the ordeal it had to pass through in Congress:

"The two great countries of which I have spoken, [England and America,] are now ringing with the praises of Cyrus W. Field, who chiefly has brought this great enterprise to its glorious and beneficent consummation. You have never heard his story; let me give you a few points in it, as a lesson that there is no condition of life in which a man, endowed with native genius, a benevolent spirit, and a courageous patience, may not become a benefactor of nations and of mankind."

After speaking of the efforts by which this New York merchant "brought into being an association of Americans and Englishmen, which contributed from surplus wealth the capital necessary as a basis for the enterprise"; he adds:

"It remained to engage the consent and the activity of the Governments of Great Britain and the United States. That was all that remained. Such consent and activity on the part of some one great nation of Europe was all that remained needful for Columbus when he stood ready to bring a new continent forward as a theatre of the world's civilization. But in each case that effort was the most difficult of all. Cyrus W. Field, by assiduity and patience, first secured consent and conditional engagement on the part of Great Britain, and then, less than two years ago, he repaired to Washington. The President and Secretary of State individually favored his proposition; but the jealousies of parties and sections in Congress forbade them to lend it their official sanction and patronage. He appealed to me. I drew the necessary bill. With the generous aid of others, Northern Representatives, and the indispensable aid of the late Thomas J. Rusk, a Senator from Texas, that bill, after a severe contest and long delay, was carried through the Senate of the United States by the majority, if I remember rightly, of one vote, and escaped defeat in the House of Representatives with equal difficulty. I have said the aid of Mr. Rusk was indispensable. If any one has wondered why I, an extreme Northern man, loved and lamented Thomas J. Rusk, an equally extreme Southern man, he has here an explanation. There was no good thing which, as it seemed to me, I could not do in Congress with his aid. When he died, it seemed to me that no good thing could be done by any one. Such was the position of Cyrus W. Field at that stage of the great enterprise. But, thus at last fortified with capital derived from New York and London, and with the navies of Great Britain and the United States at his command, he has, after trials that would have discouraged any other than a true discoverer, brought the great work to a felicitous consummation. And now the Queen of Great Britain and the President of the United States stand waiting his permission to speak, and ready to speak at his bidding; and the people of these two great countries await only the signal from him to rush into a fraternal embrace which will prove the oblivion of ages of suspicion, of jealousies and of anger."

Mr. Seward might well refer with pride to the part he took in sustaining this enterprise. He was from the beginning its firmest supporter. The bill was introduced into the Senate by him, and was carried through

SEEKING AID FROM CONGRESS.   99

mainly by his influence, seconded by Mr. Rusk, Mr. Douglas, and one or two others. It was introduced on the ninth of January, and came up for consideration on the twenty-first. Its friends had hoped that it might pass with entire unanimity. But such was the opposition, that the discussion lasted two days. The report shows that it was a subject of animated and almost angry debate, which brought out the secret of the opposition to aid being given by the Government.

Probably no measure was ever introduced in Congress for the help of any commercial enterprise, that some member, imagining that it was to benefit a particular section, did not object that it was "unconstitutional"! This objection was well answered in this case by Mr. Benjamin, of Louisiana, who asked:

"If we have a right to hire a warehouse at Port Mahon, in the Mediterranean, for storing naval stores, have we not a right to hire a company to carry our messages? I should as soon think of questioning the constitutional power of the Government to pay freight to a vessel for carrying its mail-bags across the ocean, as to pay a telegraph company a certain sum per annum for conveying its messages by the use of the electric telegraph."

This touched the precise ground on which the appropriation was asked. In their memorial to the President, the Company had said: "Such a contract will, we suppose, fall within the provisions of the Constitution in regard to postal arrangements, of which this is only a new and improved form."

Mr. Bayard, of Delaware, explained in the same terms the nature of the proposed agreement:

"It is a mail operation. It is a Post-Office arrangement. It is for the transmission of intelligence, and that is what I understood to be the function of the Post-Office Department. I hold it, therefore, to be as legitimately within the proper powers of the Government, as the employing of a stage-coach, or a steam-car, or a ship, to transport the mails, either to foreign countries, or to different portions of our own country."

Of course, as in all appropriations of money, the question of expense had to be considered, and here there were not wanting some to cry out against the extravagance of paying seventy thousand dollars a year! We had not then got used to the colossal expenditures of war, when we grew familiar with paying three millions a day! Seventy thousand dollars seemed a great sum; but Mr. Bayard in reply reminded them that England then paid nine hundred thousand dollars a year for the transportation of the mails between the United States and England; and argued that it was a very small amount for the great service rendered. He said:

"We have sent out ships to make explorations and observations in the Red Sea and in South America; we sent one or two expensive expeditions to Japan, and published at great cost some elegant books narrating their exploits. The expense even in ships alone, in that instance, was at the rate of twenty to one here, but no cry of economy was then raised." "I look upon this proposition solely as a business measure; in that point of view I believe the Government will obtain more service for the amount of money, than by any other contract that we have ever made, or now can make, for the transmission of intelligence."

As to the expense of furnishing a ship of war to assist in laying the cable, Mr. Douglas asked:

"Will it cost anything to furnish the use of one of our steamships? They are idle. We have no practical use for them at present. They are in commission. They have their coal on board, and their full armament. They will be rendering no service to us if they are not engaged in this work. If there was nothing more than a question of national pride involved, I would gladly furnish the use of an American ship for that purpose. England tenders one of her national vessels, and why should we not tender one also? It costs England nothing, and it costs us nothing."

Mr. Rusk made the same point, in arguing that ships might be sent to assist in laying the cable, giving this homely but sufficient reason: "I think that is better than to keep them rotting at the navy-yards, with the officers frollicking on shore."

Mr. Douglas urged still further:

"American citizens have commenced this enterprise. The honor and the glory of the achievement, if successful, will be due to American genius and American daring. Why should the American Government be so penurious—I do not know that that is the proper word, for it costs nothing—why should we be actuated by so illiberal a spirit as to refuse the use of one of our steamships to convey the wire when it does not cost one farthing to the Treasury of the United States?"

But behind all these objections of expense and of want of constitutional power, was one greater than all, and that was England! The real animus of the opposition was a fear of giving some advantage to Great Britain. This has always been sufficient to excite the hostility of a certain class of politicians. No matter what the subject of the proposed coöperation, if it were purely a scientific expedition, they were sure England was going to profit by it to our injury. So now there were those who felt that in this submarine cable England was literally crawling under the sea to get some advantage of the United States!

This jealousy and hostility spoke loudest from the mouths of Southerners. It is noteworthy that men who, in less than five years after, were figuring abroad, courting foreign influence against their own country, were then fiercest in denunciation of England. Mason and Slidell voted together against the bill. Butler, of South Carolina, was very bitter in his opposition—saying, with a sneer, that "this was simply a mail service under the surveillance of Great Britain"—and so was Hunter, of Virginia; while Jones, of Tennessee, bursting with patriotism, found a sufficient reason for his opposition, in that "he did not want anything to do with England or Englishmen!"

But it should be said in justice, that to this general hostility of the South there were some exceptions. Benjamin, of Loüisiana, gave the bill an earnest support; so did Mallory, of Florida, Chairman of the Naval Committee; and especially that noble Southerner, Rusk, of Texas, "with whose aid," as Mr. Seward said, "it seemed that there was no good thing which he could not do in Congress." Mr. Rusk declared that he regarded it as "the great enterprise of the age," and expressed his surprise at the very moderate subsidy asked for, only seventy thousand dollars a year, saying that, "with a reasonable prospect of success in an enterprise, calculated to produce such beneficial results, he should be willing to vote two hundred thousand dollars."

But with the majority of Southern Senators, there was a repugnance to acting in concert with England, which could not be overcome. They argued that this was not truly a line between England and the United States, but between England and her own colonies—a line of which she alone was to reap the benefit. *Both its termini were in the British possessions.* In the event of war this would give a tremendous advantage to the power holding both ends of the line. All the speakers harped on this string; and it may be worth a page or two to see how this was met and answered. When Mr. Hunter, of Virginia, asked, "What security are we to have that in time of war we shall have the use of the telegraph as well as the British Government?" Mr. Seward answered:

"It appears not to have been contemplated by the British Government that there would ever be any interruption of the amicable relations between the two countries. Therefore nothing was proposed in their contract for the contingency of war.

"That the two termini are both in the British dominions is true; but it is equally true that there is no other terminus on this continent where it is practicable to make that communication except in the British dominions. We have no dominions on the other side of the Atlantic Ocean. There is no other route known on which the telegraphic wire could be drawn through the ocean so as to find a proper resting-place or anchorage except this. The distance on this route is seventeen hundred miles. It is not even known that the telegraphic wire will carry the fluid with sufficient strength to communicate across those seventeen hundred miles. That is yet a scientific experiment, and the Company are prepared to make it.

"In regard to war, all the danger is this: There is a hazard of war at some future time, and whatever arrangements we might make, war would break them up. No treaty would save us. My own hope is, that after the telegraphic wire is once laid, there will be no more war between the United States and Great Britain. I believe that whenever such a connection as this shall be made, we diminish the chances of war, and diminish them in such a degree, that it is not necessary to take them into consideration at the present moment.

"Let us see where we are. What shall we gain by refusing to enter into this agreement? If we do not make it, the British Government has only to add ten thousand pounds sterling more annually, and they have the whole monopoly of this wire, without any stipulation whatever— not only in war but in peace. If we make this contract with the Company, we at least secure the benefit of it in time of peace, and we postpone and delay the dangers of war. If there shall ever be war, it would abrogate all treaties that can be made in regard to this subject, unless it be true, as the honorable Senator from Virginia thinks, that treaties can be made which will be regarded as obligatory by nations in time of war. If so, we have all the advantages in time of peace, for the purpose of making such treaties hereafter, without the least reason to infer that there would be any reluctance on the part of the British Government to enter into that negotiation with us, if we should desire to do so. The British Government, if it had such a disposition as the honorable Senator supposes, would certainly have proposed to monopolize all this telegraphic line, instead of proposing to divide it." *

* It is worthy of notice, that when the Bill granting a charter to the Atlantic Telegraph Company was offered in the British Parliament, at least one nobleman found fault with it on this very ground, that it gave away important advantages which properly belonged to England, and which she ought to reserve to herself:

"In the House of Lords, on the twentieth of July, 1857, on the motion for the third reading of the Telegraph Company's bill,

"Lord Redesdale called attention to the fact that, although the ter-

Mr. Hale spoke in the same strain:

"It seems to me that the war spirit and the contingencies of war are brought in a little too often upon matters of legislation which have no necessary connection with them. If we are to be governed by considerations of that sort, they would paralyze all improvements; they would stop the great appropriations for commerce; they would at once neutralize that policy which sets our ocean steamers afloat. Nobody pretends that the intercourse which is kept up between Great Britain and this country by our ocean steamers would be continued in time of war; nor the communication with France or other nations.

"If we are deterred for that reason, we shall be pursuing a policy that will paralyze improvements on those parts of the coast which lie contiguous to the lakes. The city of Detroit will have to be abandoned, beautiful and progressive

mini of the proposed telegraph were both in her Majesty's dominions, namely, in Ireland and Newfoundland, the American Government were to enjoy the same priority as the British Government with regard to the transmission of messages. It was said that this equal right was owing to the fact that a joint guarantee had been given by the two Governments. *He thought, however, it would have been far better policy on the part of her Majesty's Government if they had either undertaken the whole guarantee themselves, and thus had obtained free and sole control over the connecting line of telegraph, or had invited our own colonies to participate in that guarantee, rather than have allowed a foreign government to join in making it.* At the same time, if the clause in question had the sanction of her Majesty's ministry, it was not his intention to object to it.

"Earl Granville said this telegraph was intended to connect two great countries, and, as the two Governments had gone hand in hand with regard to the guarantee, it seemed only reasonable that both should have the same rights as to transmitting messages.

"The bill was then read a third time and passed."

SEEKING AID FROM CONGRESS. 107

as it is, because in time of war the mansions of her citizens there lie within the range of British guns.

"What will the suspension bridge at Niagara be good for in a time of war? If the British cut off their end of it, our end will not be worth much. I believe that among the things which will bind us together in peace, this telegraphic wire will be one of the most potent. It will bind the two countries together literally with cords of iron that will hold us in the bonds of peace. I repudiate entirely the policy which refuses to adopt it, because in time of war it may be interrupted. Such a policy as that would drive us back to a state of barbarism. It would destroy the spirit of progress; it would retard improvement; it would paralyze all the advances which are making us a more civilized, and a more informed and a better people than the one which preceded us."

Mr. Douglas cut the matter short by saying:

"I am willing to vote for this bill as a peace measure, as a commercial measure—but not as a war measure; and when war comes, let us rely on our power and ability to take this end of the wire, and keep it."

Mr. Benjamin said:

"The sum of money that this Government proposes to give for the use of this telegraph will amount, in the twenty-five years, to something between £300,000 and £400,000. Now, if this be a matter of such immense importance to Great Britain—if this be the golden opportunity—and if, indeed, her control of this line be such a powerful engine, whether in war or in peace, is it not most extraordinary that she proposes to us a full share in its benefits and in its control, and allows to our Government equal rights with herself in the transmission of communications for the sum of about £300,000, to be paid in annual instalments through twenty-five years ? If this be, indeed, a very important instrumentality in behalf of Great Britain for the conduct of her commerce, the government of her possessions, or the efficient action of her troops in time of war, the £300,000 expended upon it are but as a drop in the bucket when compared with the immense resources of that empire. I think, therefore, we may as well discard from our consideration of this subject all these visions about the immense importance of the governmental aid in this matter, to be rendered under the provisions of this bill.

"Mr. President, let us not always be thinking of war ; let us be using means to preserve peace. The amount that would be expended by this Government in six months' war with Great Britain, would far exceed every thing that we shall have to pay for the use of this telegraphic line for the entire twenty-five years of the contract; and do you not believe that this instrumentality will be sufficiently efficient to bind together the peace, the commerce, and the interests of the two countries, so as even to defer a war for six months or twelve months, if one should ever become inevitable, beyond the period at which it would otherwise occur ? If it does that, it will in six or eight or nine months repay the expenditures of twenty-five years.

"Again, Sir, I say, if Great Britain wants it for war, she will put it there at her own expense. It is not three hundred thousand pounds, or four hundred thousand pounds, that will arrest her. If, on the contrary, this be useful to commerce—useful in an eminent degree—useful for the preservation of peace, then I confess I feel some pride that my country should aid in establishing it. I confess 1 feel a glow of something like pride that I belong to the great human family when I see these triumphs of science, by which mind is brought into instant communication with mind across the intervening oceans, which, to our unenlightened forefathers, seemed placed there by Providence as an eternal barrier to communication between man and man. Now, Sir, we speak from minute to minute. Scarcely can a gun be fired in war on the European shore ere its echoes will reverberate among our own mountains, and be heard by every citizen in the land. All this is a triumph of science—of American genius, and I for one feel proud of it, and feel desirous of sustaining and promoting it."

Mr. Douglas said:

"Our policy is essentially a policy of peace. We want peace with the whole world, above all other considerations. There never has been a time in the history of this Republic, when peace was more essential to our prosperity, to our advancement, and to our progress, than it is now. We have made great progress in time of peace—an almost inconceivable progress since the last war with Great Britain. Twenty-five years more of peace will put us far in advance of any other nation on earth."

It was fit that Mr. Seward, who introduced the bill, and opened the debate, should close in words that now seem prophetic, and show the large wisdom, looking before and after, of this eminent statesman:

"There was an American citizen who, in the year 1770, or thereabout, indicated to this country, to Great Britain, and to the world, the use of the lightning for the purposes of communication of intelligence, and that was Dr. Franklin. I am sure that there is not only no member of the Senate, but no American citizen, however humble, who would be willing to have struck out from the achievements of American invention this great discovery of the lightning as an agent for the uses of human society.

"The suggestion made by that distinguished and illustrious American was followed up some fifty years afterward by another suggestion and another indication from another American, and that was Mr. Samuel F. B. Morse, who indicated to the American Government the means by which the lightning could be made to write, and by which the telegraphic wires could be made to supply the place of wind and steam for carrying intelligence.

"We have followed out the suggestions of these eminent Americans hitherto, and I am sure at a very small cost. The Government of the United States appropriated $40,000 to test the practicability of Morse's suggestion; the $40,000 thus expended established its practicability and its use. Now, there is no person on the face of the globe who can measure the price at which, if a reasonable man, he would be willing to strike from the world the use of the magnetic telegraph as a means of communication between different portions of the same country. This great invention is now to be brought into its further, wider, and broader use—the use by the general society of nations, international use, the use of the society of mankind. Its benefits are large—just in proportion to the extent and scope of its operation. They are not merely benefits to the Government, but they are

SEEKING AID FROM CONGRESS. 111

benefits to the citizens and subjects of all nations and of all States.

"I might enlarge further on this subject, but I forbear to do so, because I know that at some future time I shall come across the record of what I have said to-day. I know that then what I have said to-day, by way of anticipation, will fall so far short of the reality of benefits which individuals, States, and nations will have derived from this great enterprise, that I shall not reflect upon it without disappointment and mortification."

After such arguments, it should seem that there could be but one opinion, and yet the bill passed the Senate by only *one* majority! It also had to run the gauntlet of the House of Representatives, where it encountered the same hostility. But at length it got through, and was signed by President Pierce on the third of March, the day before he went out of office. Thus it became a law.

CHAPTER VIII.

THE EXPEDITION OF 1857.

SCARCELY was the business with the American Government completed, before Mr. Field was recalled to England. Once more upon the waves, he forgot the long delay and the vexatious opposition which he left behind—the fogs of Newfoundland, and the denser fogs of Washington. He was bound for England, and there at least the work did not stand still. All winter long the wheels of the machinery had kept in motion. The cable was uncoiling its mighty folds to a length sufficient to span the Atlantic, and at last there was hope of victory.

Although the United States Government had seemed a little ungracious in its delay, it yet rendered, this year and the next, most important service. Already it had prepared the way, by the deep-sea soundings, which it was the first to take across the Atlantic. It now rendered additional and substantial aid in lending to this enterprise the two finest ships in the American navy—the Niagara and the Susquehanna. The former was built some dozen years before by George Steers— a name celebrated among our marine architects as the constructor of the famous yacht America, that "race-horse of the sea," which had crossed the Atlantic, and carried off the prize in the British Channel from the yachts of England—and was designed to be a model of naval architecture. She was the largest steam-frigate in the world, exceeding in tonnage the heaviest line-of-battle ship in the English navy, and yet so finely modelled that, propelled only by a screw, she could make ten or twelve miles an hour. Notwithstanding her bulk, she was intended to carry but twelve guns—being one of the first ships in our navy to substitute a few heavy Dahlgrens for half a dozen times as many fifty-six-pounders. This was the beginning of that revolution in naval warfare, which was carried to such extent in the Monitors and other ironclads introduced in our civil war. Each gun weighed fourteen tons—requiring a crew of twenty-five men to wield it—and threw a shell of one hundred and thirty pounds a distance of three miles. One or two broadsides from such a deck would sink an old-fashioned seventy-four, or even a ninety or hundred-gun ship.

But as the Niagara was now to go on an errand of peace, this formidable armament was not taken on board. She was built with what is known as a flush deck, clear from stem to stern, and being without her guns, was left free for the more peaceful burden that she was to bear. When the orders were received from Washington, she was lying at the Brooklyn

Navy-Yard, but began immediately to prepare for her expedition. Bulkheads were knocked down, above and below, to make room for the huge monster of the deep that was to be coiled within her sides. These preparations occupied four or five weeks. On the twenty-second of April, she made a trial trip down the bay, and two days after sailed for England, in command of Captain William L. Hudson, one of the oldest and best officers in our navy, who, to his past services to his country, was now to add another in the expeditions of this and the following year. He had with him as Chief Engineer Mr. William E. Everett, whose mechanical genius proved so important in constructing the paying-out machinery.

Besides the regular ship's crew, no one was received on board except Mr. Field and Professor Morse, who went as the electrician of the Newfoundland Company; and two officers of the Russian navy—Captain Schwartz and Lieutenant Kolobnin—who were permitted by our Government, as an act of national courtesy, to go out to witness the great experiment. The regulations of the navy did not admit correspondents of the press; but Professor Morse was permitted to take a secretary, and chose Mr. Mullaly, who reported for the New York Herald, and who had thus an opportunity to witness all the preparations on land and sea, and to furnish those minute and detailed accounts of the several expeditions, which contribute some important chapters in the history of this enterprise.

The Niagara arrived out on the fourteenth of May, and cast anchor off Gravesend, about twenty-five miles below London. As it was the first time—at least for many years—that an American ship of war had appeared in the Thames, this fact, with her fine proportions and the object for which she came, attracted a crowd of visitors. Every day, from morning to night, a fleet of boats was around her, and men and women thronged over her sides. Everybody was welcome. All were received with the utmost courtesy, and allowed access to all parts of the ship. Among these were many visitors of distinction. Here came Lady Franklin to thank the generous nation that had sent two expeditions to recover her husband lost amid Polar seas. She was, of course, the object of general attention and respectful sympathy.

While lying in the Thames, the Agamemnon, that was to take the other half of the cable, passed up the river. This was a historical ship, having borne the flag of the British admiral at the bombardment of Sebastopol, and distinguished herself by steaming up within a few hundred yards of the guns of the fortress. After passing through the fires of that terrible day, she was justly an object of pride to Britons, whose hearts swelled as they saw this oak-ribbed leviathan, that had come " out of the gates of death, out of the jaws of hell," now preparing to take part in achievements of peace, not less glorious than those of war. She was under command of Captain Noddal, of the Royal Navy.

As the Agamemnon came up the river in grand style, she recognized the Niagara lying off Gravesend, and manning her yards, gave her a succession of those English hurras so stirring to the blood, when heard on land or sea, to which our tars replied with lusty American cheers. It was pleasant to observe, from this time, the hearty good-will that existed between the officers and crews of the two ships, who in their exertions for the common object, were animated only by a generous rivalry.

A few days after, the Niagara was joined by the Susquehanna, Captain Sands, which had been ordered from the Mediterranean to take part also in the expedition. She was a fit companion ship, being the largest side-wheel steamer in our navy, as the other was the largest propeller. Both together, they were worthy representatives of the American navy.

When the Niagara arrived in the Thames, it was supposed she would take on board her half of the cable from the manufactory of Glass, Elliot & Co., at Greenwich; but on account of her great length, it was difficult to bring her up alongside the wharf in front of the works. This was therefore left to the

Agamemnon, while the Niagara was ordered around to Liverpool, to take the other half from the works of Newall & Co., at Birkenhead, opposite that city. Accordingly she left Gravesend on the fifth of June, and reached Portsmouth the next day, where she remained a fortnight, to have some further alterations to fit her to receive the cable. Although she had been already pretty well "scooped out," fore and aft, the cry was still for room. Officers had to shift for themselves, as their quarters were swept away to make a wider berth for their iron guest. But all submitted with excellent grace. Like true sailors, they took it gayly as if they were only clearing the decks for battle. Among other alterations for safety, was a framework or cage of iron, which was put over the stern of the ship, to keep the cable from getting entangled in the screw. As soon as these were completed, the Niagara left for Liverpool, and on the twenty-second of June cast anchor in the Mersey. Here she attracted as much attention as in the Thames, being crowded with visitors during the week; and on Sundays, when none were received on board, the river boats sought to gratify public curiosity by sailing round her. The officers of the ship were objects of constant hospitality, both from private citizens and from the public authorities. The Mayor of Liverpool gave them a dinner, the Chamber of Commerce another, while the Americans in Liverpool entertained them on the fourth of

July—the first public celebration of our national anniversary ever had in that city.

But while these festivities were kept up on shore, hard work was done on board the ship. To coil thirteen hundred miles of cable was an immense undertaking. Yet it was all done by the sailors themselves. No compulsion was used, and none was needed. No sooner was there a call for volunteers, than men stepped forward in greater numbers than could be employed. Out of these were chosen one hundred and twenty stalwart fellows, who were divided into two gangs of sixty men, and each gang into watches of thirty, which relieved each other, and all went to work with such enthusiasm, that in three weeks the herculean task was completed. The event was celebrated by a final dinner given by the shareholders of the Atlantic Telegraph Company in Liverpool to Captain Hudson and Captain Sands of the Susquehanna, whose arrival in the Mersey enabled them to extend their hospitalities to the officers of both ships.

While the Niagara was thus doing her part, the same scene was repeated on board the Agamemnon, which was still lying in the Thames. There the work was completed about the same day, and the occasion duly honored by a scene as unique as it was beautiful. Says the London Times of July twenty-fourth:

"All the details connected with the manufacture and stowage of the cable are now completed, and the conclusion of the arduous labor was celebrated yesterday with high festivity and rejoicing. All the artisans who have been engaged upon the great work, with their wives and families, a large party of the officers, with the sailors from the Agamemnon, and a number of distinguished scientific visitors, were entertained upon this occasion at a kind of *fête champêtre* at Belvidere House, the seat of Sir Culling Eardley, near Erith. The festival was held in the beautiful park which had been obligingly opened by Sir Culling Eardley for the purpose. Although in no way personally interested in the project, the honorable baronet has all along evinced the liveliest sympathy with the undertaking, and himself proposed to have the completion of the work celebrated in his picturesque grounds. The manufacturers, fired with generous emulation, erected spacious tents on the lawn, and provided a magnificent banquet for the guests, and a substantial one for the sailors of the Agamemnon and the artificers who had been employed in the construction of the cable. By an admirable arrangement, the guests were accommodated at a vast semi-circular table, which ran round the whole pavilion, while the sailors and workmen sat at a number of long tables arranged at right angles with the chord, so that the general effect was that all dined together, while at the same time sufficient distinction was preserved to satisfy the most fastidious. The three centre tables were occupied by the crew of the Agamemnon, a fine, active body of young men, who paid the greatest attention to the speeches, and drank all the toasts with an admirable punctuality, at least so long as their three pints of beer per man lasted ; but we regret to add that, what with the heat of the day and the enthusiasm of Jack in the cause of science, the mugs were all empty long before the chairman's list of toasts had been gone through. Next in interest to the sailors were the workmen and their wives and babies, all being permitted to assist at the great occasion. The latter, it is true, sometimes squalled at an affecting peroration, but that rather improved the effect than otherwise, and the presence of these little ones only marked the genuine good feeling of the employers, who had thus invited not only their workmen, but their workmen's families to the feast. It was a momentary return to the old patriarchal times, and every one present seemed delighted with the experiment."

Speeches were made by Sir Culling Eardley, by Mr. Cardwell, of the House of Commons, Mr. Brooking, one of the Directors, by Professor Morse, and others. Mr. Field read a letter from President Buchanan, saying that he should feel honored if the first message should be one from Queen Victoria to himself, and that he "would endeavor to answer it in a spirit and manner becoming a great occasion."

Thus, labor and feasting being ended, the Niagara and the Susquehanna left Liverpool the latter part of July and steamed down St. George's Channel to Queenstown, which was to be the rendezvous of the telegraphic squadron, where they were joined by the Agamemnon and the Leopard, which was to be her consort. The former, as she entered the harbor, came to anchor about a third of a mile from the Niagara. The presence of the two ships which had the cable on board, gave an opportunity which the electricians had desired to test its integrity. Accordingly one end of each cable was carried to the opposite ship, and so joined as to form a continuous length of twenty-five hundred miles, both ends of which were on board the Agamemnon. One end was then connected with the apparatus for transmitting the electric current, and on a sensitive galvanometer being attached to the other end, the whole cable was tested from end to end, and found to be perfect. These experiments were continued for two days with the same result. This inspired fresh hopes for the success of the expedition, and in high spirits they bore away for the harbor of Valentia.

It had been for some time a matter of discussion, where they should begin to lay the cable, whether from the coast of Ireland, or in mid-ocean, the two ships making the junction there, and dropping it to the bottom of the sea, and then parting, one to the east and the other to the west, till they landed their ends on the opposite shores of the Atlantic. This was the plan adopted the following year, and which finally proved successful. It was the one preferred by the engineers now, but the electricians favored the other course, and their counsel prevailed. It was therefore decided to submerge the whole cable in a continuous line from Valentia Bay to Newfoundland. The Niagara was to lay the first half from Ireland to the middle of the Atlantic; the end would then be joined to the other half on board the Agamemnon, which would take it on to the coast of Newfoundland.

During the whole process the four vessels were to remain together and give whatever assistance was required. While it was being laid down, messages were to be sent back to Valentia, reporting each day's progress.

As might be supposed, the mustering of such a fleet of ships, and the busy note of preparation which had been heard for weeks, produced a great sensation in this remote part of Ireland. The people from far and near, gathered on the hills and looked on in silent wonder.

To add to the dignity of the occasion, the Lord Lieutenant came down from Dublin to witness the departure of the expedition. No one could have been better fitted to represent his own country, and to command audience from ours. The Earl of Carlisle—better known among us as Lord Morpeth—had travelled in the United States a few years before, and shown himself one of the most intelligent and liberal foreigners that have visited America. No representative of England could on that day have stood upon the shores of Ireland, and stretched out his hand to his kindred beyond the sea with more assurance that his greeting would be warmly responded to. And never did one speak more aptly words of wisdom and of peace. We read them still with admiration for their beauty and their eloquence, and with an interest more tender but more sad, that this great and good man—

the true friend of his own country and of ours—has gone to his grave. To quote his own words is the best tribute to his memory, and will do more than any eulogy to keep it fresh and green in the hearts of Americans. On his arrival at Valentia, he was entertained by the Knight of Kerry at one of those public breakfasts so much in fashion in England, at which in response to a toast in his honor, after making his personal acknowledgments, he said:

"I believe, as your worthy chairman has already hinted, that I am probably the first Lieutenant of Ireland who ever appeared upon this lovely strand. At all events, no Lord Lieutenant could have come amongst you on an occasion like the present. Amidst all the pride and the stirring hopes which cluster around the work of this week, we ought still to remember that we must speak with the modesty of those who begin and not of those who close an experiment, and it behooves us to remember that the pathway to great achievements has frequently to be hewn out amidst risks and difficulties, and that preliminary failure is even the law and condition of the ultimate success. Therefore, whatever disappointments may possibly be in store, I must yet insinuate to you that in a cause like this it would be criminal to feel discouragement. In the very design and endeavor to establish the Atlantic Telegraph there is almost enough of glory. It is true if it be only an attempt there would not be quite enough of profit. I hope that will come, too ; but there is enough of public spirit, of love for science, for our country, for the human race, almost to suffice in themselves. However, upon this rocky frontlet of Ireland, at all events, to-day we will presume upon success. We are about, either by this sundown or by to-morrow's dawn, to establish a new material link between the Old World and the New. Moral links there have been—links of race, links of commerce, links of friendship, links of literature, links of glory ; but this, our new link, instead of superseding and supplanting the old ones, is to give a life and an intensity which they never had before. Highly as I value the reputations of those who have conceived, and those who have contributed to carry out this bright design—and I wish that so many of them had not been unavoidably prevented from being amongst us at this moment*—highly as I estimate their reputation, yet I do not compliment them with the idea that they are to efface or dim the glory of that Columbus, who, when the large vessels in the harbor of Cork yesterday weighed their anchors, did so on that very day three hundred and sixty-five years ago—it would have been called in Hebrew writ a year of years—and set sail upon his glorious enterprise of discovery. They, I say, will not dim or efface his glory, but they are now giving the last finish and consummation to his work. Hitherto the inhabitants of the two worlds have associated perhaps in the chilling atmosphere of distance with each other—a sort of bowing distance ; but now we can be hand to hand, grasp to grasp, pulse to pulse. The link, which is now to connect us, like the insect in the immortal couplet of our poet :

> While exquisitely fine,
> Feels at each thread and lives along the line.

And we may feel, gentlemen of Ireland, of England, and of America, that we may take our stand here upon the extreme

* Mr. Field was detained by illness at Valentia, and several of the ships had not arrived.

rocky edge of our beloved Ireland; we may, as it were, leave in our rear behind us the wars, the strifes, and the bloodshed of the elder Europe, and of the elder Asia; and we may pledge ourselves, weak as our agency may be, imperfect as our powers may be, inadequate in strict diplomatic form as our credentials may be, yet, in the face of the unparalleled circumstances, of the place and the hour, in the immediate neighborhood of the mighty vessels whose appearance may be beautiful upon the waters, even as are the feet upon the mountains of those who preach the Gospel of peace—as an homage due to that serene science which often affords higher and holier lessons of harmony and good will than the wayward passions of man are always apt to learn—in the face and in the strength of such circumstances, let us pledge ourselves to eternal peace between the Old World and the New."

While these greetings were exchanged on shore, only the smaller vessels of the squadron had arrived. But in a few hours the great hulls of the Niagara and the Agamemnon, followed by the Leopard and the Susquehanna, were seen in the horizon, and soon they all cast anchor in the bay. As the sun went down in the west, shining still on the other hemisphere which they were going to seek, its last rays fell on an expedition more suggestive and hopeful than any since that of Columbus from the shores of Spain, and upon navigators not unworthy to be his followers.

The whole squadron was now assembled, and made gallant array. There were present in the little harbor of Valentia seven ships—the stately Niagara, which was to lay the half of the cable from Ireland, and her consort, the Susquehanna, riding by her side; while floating the flag of England, were the Agamemnon, which was appointed to lay the cable on the American side, and her consort, the Leopard. Beside these high-decked ships of war, the steamer Advice had come round to give, not merely advice but lusty help in landing the cable at Valentia; and the little steamer Willing Mind, with a zeal worthy of her name, was flying back and forth between ship and shore, lending a hand wherever there was work to be done; and the Cyclops, under the experienced command of Captain Dayman, who had made the deep-sea soundings across the Atlantic only the month before, here joined the squadron to lead the way across the deep. This made five English ships, with but two American; but to keep up our part, there were two more steamers on the other side of the sea, the Arctic, under Lieutenant Berryman, and the Company's steamer Victoria, to watch for the coming of the fleet off the coast of Newfoundland, and help in landing the cable on the shores of the New World.

It was now Tuesday evening, the fourth of August, too late to undertake the landing that night, but preparations were at once begun for it the next morning. Said the correspondent of the Liverpool Post:

" The ships were visited in the course of the evening by the Directors and others interested in the great undertaking, and arrangements were immediately commenced on board the Niagara for paying out the shore rope for conveyance to the mainland. These arrangements were fully perfected by Wednesday morning; but for some hours the state of the weather rendered it doubtful whether operations could be safely proceeded with. Toward the afternoon the breeze calmed down, and at two o'clock it was decided that an effort should be made to land the cable at once. The process of uncoiling into the small boats commenced at half-past two, and the scene at this period was grand and exciting in the highest degree.

" Valentia Bay was studded with innumerable small craft, decked with the gayest bunting—small boats flitted hither and thither, their occupants cheering enthusiastically as the work successfully progressed. The cable-boats were managed by the sailors of the Niagara and Susquehanna, and it was a well-designed compliment, and indicative of the future fraternization of the nations, that the shore rope was arranged to be presented at this side the Atlantic to the representative of the Queen, by the officers and men of the United States navy, and that at the other side the British officers and sailors should make a similar presentation to the President of the Great Republic.

" From the main land the operations were watched with intense interest. For several hours the Lord Lieutenant stood on the beach, surrounded by his staff and the directors of the railway and telegraph companies, waiting the arrival of the cable, and when at length the American sailors jumped through the surge with the hawser to which it was attached, his Excellency was among the first to lay hold of it and pull it lustily to the shore. Indeed every one present seemed desirous of having a hand in the great work; and never before perhaps were there so many willing assistants, at 'the long pull, the strong pull, and the pull all together.'

"At half-past seven o'clock the cable was hauled on shore, and formal presentation was made of it to the Lord Lieutenant by Captain Pennock, of the Niagara; his Excellency expressing a hope that the work so well begun would be carried to a satisfactory completion."

The wire having been secured to a house on the beach, the Reverend Mr. Day, of Kenmore, advanced and offered the following prayer :

"O Eternal Lord God, who alone spreadest out the heavens, and rulest the raging of the sea; who hast compassed the water with bounds, till day and night come to an end; and whom the winds and the sea obey; look down in mercy, we beseech thee, upon us thy servants, who now approach the throne of grace; and let our prayer ascend before thee with acceptance. Thou hast commanded and encouraged us, in all our ways, to acknowledge thee, and to commit our works to thee; and thou hast graciously promised to direct our paths, and to prosper our handiwork. We desire now to look up to thee; and believing that without thy help and blessing, nothing can prosper or succeed, we humbly commit this work, and all who are engaged in it, to thy care and guidance. Let it please thee to grant to us thy servants wisdom and power, to complete what we have been led by thy Providence to undertake; that being begun and carried on in the spirit of prayer, and in dependence upon thee, it may tend to thy glory: and to the good of all nations, by promoting the increase of unity, peace, and concord.

"Overrule, we pray thee, every obstacle, and remove every difficulty which would prevent us from succeeding in this important undertaking. Control the winds and the sea by thy Almighty power, and grant us such favorable weather that we may be enabled to lay the Cable safely and effectually. And may thy hand of power and mercy be so acknowledged by all, that the language of every heart may be, 'Not unto us, O Lord, not unto us, but unto thy name give glory,' that so thy name may be hallowed and magnified in us and by us.

"Finally, we beseech thee to implant within us a spirit of humility and childlike dependence upon thee; and teach us to feel as well as to say, 'If the Lord will, we shall do this or that.'

"Hear us, O Lord, and answer us in these our petitions, according to thy precious promise, for Jesus Christ's sake. Amen."

The Lord Lieutenant then spoke once more—words that amid such a scene and at such an hour, sank into all hearts :

"My American, English, and Irish friends, I feel at such a moment as this that no language of mine can be becoming except that of prayer and praise. However, it is allowable to any human lips, though they have not been specially qualified for the office, to raise the ascription of 'Glory to God in the highest; on earth peace, good-will to men.' That, I believe, is the spirit in which this great work has been undertaken; and it is this reflection that encourages me to feel confident hopes in its final success. I believe that the great work now so happily begun will accomplish many great and noble purposes of trade, of national policy, and of empire. But there is only one view in which I will present it to those whom I have the pleasure to address. You are aware—you must know, some of you, from your own experience—that many of your dear friends and near relatives have left their native land to receive hospitable shelter in America. Well, then, I do not expect that all of you can understand the wondrous mechanism by which this great undertaking is to be carried on. But this, I think, you all of you understand. If you wished to communicate some piece of intelligence straightway to your relatives across the wide world of waters—if you wished to tell those whom you know it would interest in their heart of hearts, of a birth, or a marriage, or, alas, a death, among you, the little cord, which we have now hauled up to the shore, will impart that tidings quicker than the flash of the lightning. Let us indeed hope, let us pray that the hopes of those who have set on foot this great design, may be rewarded by its entire success; and let us hope, further, that this Atlantic Cable will, in all future time, serve as an emblem of that strong cord of love which I trust will always unite the British islands to the great continent of America. And you will join me in my fervent wish that the Giver of all good, who has enabled some of his servants to discern so much of the working of the mighty laws by which he fills the universe, will further so bless this wonderful work, as to make it even more to serve the high purpose of the good of man, and tend to his great glory. And now, all my friends, as there can be no project or undertaking which ought not to receive the approbation and applause of the people, will you join with me in giving three hearty cheers for it? [Loud cheering.] Three cheers are not enough for me—they are what we give on common occasions—and as it is for the success of the Atlantic Telegraph Cable, I must have at least one dozen cheers. [Loud and protracted cheering.]"

Mr. Brooking, the Chairman of the Executive Committee of the Atlantic Telegraph Company, then expressed the thanks which all felt to the Lord Lieutenant for his presence on that occasion.

Then there were loud calls for Mr. Field. He could only answer:

"I have no words to express the feelings which fill my heart to-night—it beats with love and affection for every man, woman and child who hears me. I may say, however, that, if ever at the other side of the waters now before us, any one of you shall present himself at my door and say that he took hand or part, even by an approving smile, in our work here to-day, he shall have a true American welcome. I cannot bind myself to more, and shall merely say : 'What God has joined together, let not man put asunder.'"

Thus closed this most interesting scene. The Lord Lieutenant was obliged to return at once to the capital. He therefore left, and posted that night to Killarney, and the next day returned by special train to Dublin, leaving the ships to complete the work so happily begun.

The landing of the cable took place on Wednesday, the fifth of August, near the hour of sunset. As it was too late to proceed that evening, the ships remained at anchor till the morning. They got under weigh at an early hour, but were soon checked by an accident which detained them another day. Before they had gone five miles, the heavy shore end of the cable caught in the machinery and parted. The Niagara put back, and the cable was "underrun" the whole distance. At length the end was lifted out of the water and spliced to the gigantic coil, and as it dropped safely to the bottom of the sea, the mighty ship began to stir. At first she moved very slowly, not more than two miles an hour, to avoid the danger of accident; but the feeling that they were at last away was itself a relief. The ships were all in sight, and so near that they could hear each other's bells. The Niagara, as if knowing that she was bound for the land out of whose forests she came, bowed her head to the waves, as her prow was turned toward her native shores.

Slowly passed the hours of that day. But all went well, and the ships were moving out into the broad Atlantic. At length the sun went down in the west, and stars came out on the face of the deep. But no man slept. A thousand eyes were watching a great experiment as those who have a personal interest in the issue. All through that night, and through the anxious days and nights that followed, there was a feeling in every soul on board, as if some dear friend were at the turning-point of life or death, and they were watching beside him. There was a strange, unnatural silence in the ship. Men paced the deck with soft and muffled tread, speaking only in whispers, as if a loud voice or a heavy footfall might snap the vital cord. So much had they grown to feel for the enterprise, that the cable seemed to them like a human creature, on whose fate they hung, as if it were to decide their own destiny.

There are some who will never forget that first night at sea. Perhaps the reaction from the excitement on shore made the impression the deeper. There are moments in life when every thing comes back upon us. What memories came up in those long night hours! How many on board that ship, as they stood on the deck and watched that mysterious cord disappearing in the darkness, thought of homes beyond the sea, of absent ones, of the distant and the dead!

But no musings turn them from the work in hand. There are vigilant eyes on deck. Mr. Bright, the engineer of the Company, is there, and Mr. Everett, Mr. De Sauty, the electrician, and Professor Morse. The paying-out machinery does its work, and though it makes a constant rumble in the ship, that dull, heavy sound is music to their ears, as it tells them that all is well. If one should drop to sleep, and wake up at night, he has only to hear the sound of "the old coffee-mill," and his fears are relieved, and he goes to sleep again.

Saturday was a day of beautiful weather. The ships were getting farther away from land, and began to steam ahead at the rate of four and five miles an hour. The cable was paid out at a speed a little faster than that of the ship, to allow for any inequalities of surface on the bottom of the sea. While it was thus going overboard, communication was kept up constantly with the land. Every moment the current was passing between ship and shore. The communication was as perfect as between Liverpool and London, or Boston and New York. Not only did the electricians telegraph back to Valentia the progress they were making, but the officers on board sent messages to their friends in America, to go out by the steamers from Liverpool. The heavens seemed to smile on them that day. The coils came up from below the deck without a kink, and unwinding themselves easily, passed over the stern into the sea. Once or twice an alarm was created by the cable being thrown off the wheels. This was owing to the sheaves not being wide enough and deep enough, and being filled with tar, which hardened in the air. This was a great defect of the machinery which was remedied in the later expeditions. Still it worked well, and so long as those terrible brakes kept off their iron gripe, it might work through to the end.

All day Sunday the same favoring fortune continued; and when the officers, who could be spared from the deck, met in the cabin, and Captain Hudson read the service, it was with subdued voices and grateful hearts they responded to the prayers to Him who spreadeth out the heavens, and ruleth the raging of the sea.

On Monday they were over two hundred miles at sea. They had got far beyond the shallow waters off the coast. They had passed over the submarine mountain which figures on the charts of Dayman and Berryman, and where Mr. Bright's log gives a descent from five hundred and fifty to seventeen hundred and fifty fathoms within eight miles! Then they came to the deeper waters of the Atlantic, where the cable sank to the awful depth of two thousand fathoms. Still the iron cord buried itself in the waves, and every instant the flash of light in the darkened telegraph room told of the passage of the electric current.

But Monday evening, about nine o'clock, occurred a mysterious interruption, which staggered all on board. Suddenly the electrical continuity was lost. The cable was not broken, but it ceased to work. Here was a mystery. De Sauty tried it, and Professor Morse tried it. But neither could make it work. It seemed that all was over. The electricians gave it up, and the engineers were preparing to cut the cable, and to endeavor to wind it in, when suddenly *the electricity came back again.* This made the mystery greater than ever. It had been interrupted for two hours and a half. This was a phenomenon which has never been explained. Professor Morse was of opinion that the cable, in getting off the wheels, had been strained so as to open the gutta-percha, and thus destroy the insulation. If this be the true explanation, it would seem that on reaching the bottom the seam had closed, and thus the continuity had been restored. But it was certainly an untoward incident, which "cast ominous conjecture on the whole success," as it seemed to indicate that there were at the bottom of the sea causes which were wholly unknown and against which it was impossible to provide.

The return of the current was like life from the dead. Says Mullaly:

"The glad news was soon circulated throughout the ship, and all felt as if they had a new life. A rough, weather-beaten old sailor, who had assisted in coiling many a long mile of it on board the Niagara, and who was among the first to run to the telegraph office to have the news confirmed, said he would have given fifty dollars out of his pay to have saved that cable. 'I have watched nearly every mile of it,' he added, 'as it came over the side, and I would have given fifty dollars, poor as I am, to have saved it, although I don't expect to make any thing by it when it is laid down.' In his own simple way he expressed the feelings of every one on board, for all are as much interested in the success of the enterprise as the largest shareholder in the Company. They talked of the cable as they would of a pet child, and never was child treated with deeper solicitude than that with which the cable is watched by them. You could see the tears standing in the eyes of some as they almost cried for joy, and told their messmates that it was all right."

THE EXPEDITION OF 1857.   137

It was indeed a great relief; and though still anxious, after watching till past midnight, a few crept to their couches, to snatch an hour or two of broken sleep. But before the morning broke, the hopes thus revived were again and finally destroyed.

The cable was running out freely at the rate of six miles an hour, while the ship was advancing but about four. This was supposed to be owing to a powerful under-current. To check this waste, the engineer applied the brakes firmly, which at once stopped the machine. The effect was to bring a heavy strain on the cable that was in the water. The stern of the ship was down in the trough of the sea, and as it rose upward on the swell, the tension was too great, and the cable parted.

Instantly ran through the ship a cry of grief and dismay. She was stopped in her onward path, and in a few minutes all gathered on deck with feelings which may be imagined. One who was present wrote: "The unbidden tear started to many a manly eye. The interest taken in the enterprise by all, every one, officers and men, exceeded any thing I ever saw, and there is no wonder that there should have been so much emotion at our failure." Captain Hudson says: "It made all hands of us through the day like a household or family which had lost their dearest friend, for officers and men had been deeply interested in the success of the enterprise."

There was nothing left but to return to England. The position is very clearly stated by Mr. Field in a letter to one of his family, which shows how his own courage survived the great disaster:—

"H. M. Steamer Leopard, Thursday, August 13, 1857.

"The successful laying down of the Atlantic Telegraph Cable is put off for a short time, but its final triumph has been fully proved, by the experience that we have had since we left Valentia. My confidence was never so strong as at the present time, and I feel sure, that with God's blessing, we shall connect Europe and America with the electric cord.

"After having successfully laid—and part of the time while a heavy sea was running—three hundred and thirty-five miles of the cable, and over one hundred miles of it in water more than two miles in depth, the brakes were applied more firmly, by order of Mr. Bright, the engineer, to prevent the cable from going out so fast, and it parted.

"I retired to my state-room at a little after midnight Monday, all going on well, and at a quarter before four o'clock on Tuesday morning, the eleventh instant, I was awoke from my sleep by the cry of 'Stop her, back her!' and in a moment Mr. Bright was in my room, with the sad intelligence that the cable was broken. In as short a time as possible I was dressed, and on deck; and Captain Hudson at once signaled the other steamers that the cable had parted, and in a few moments Captain Wainwright, of the Leopard, and Captain Sands, of the Susquehanna, were on board of the Niagara.

THE EXPEDITION OF 1857.

"I requested Captain Wainwright, the commander of the English Telegraph Fleet, to order the Agamemnon to remain with the Niagara and Susquehanna, in this deep part of the Atlantic for a few days, to try certain experiments which will be of great value to us, and then sail with them back to England, and all wait at Plymouth until further orders. I further requested Captain Wainwright to order the Cyclops to sound here where the cable parted, and then steam back to Valentia, with letters from me to Dr. Whitehouse, and Mr. Saward, the secretary of the Atlantic Telegraph Company ; and that he should take me in the Leopard as soon as possible to Portsmouth.

"All of my requests were cheerfully complied with, and in a few hours the Cyclops had sounded, and found the bottom at two thousand fathoms, and was on her way back to Valentia with letters from me ; the Niagara and the Agamemnon were connected together by the cable, and engaged in trying experiments ; the Susquehanna in attendance, and the Leopard, with your affectionate ——— on board, on her way back to England.

"In my letter to Dr. Whitehouse I requested him to telegraph to London, and have a special meeting of the Directors called for twelve o'clock on Saturday, to decide whether we should have more cable made at once, and try again this season, or wait until next year.

"I shall close this letter on board, so as to have it ready to mail the moment we arrive at Portsmouth, as I wish to leave by the very next train for London, so as to be there in time to meet the Directors Saturday noon, and read them my report, which I am busy making up.

"Do not think that I feel discouraged, or am in low spirits, for I am not; and I think I can see how this accident will be of great advantage to the Atlantic Telegraph Company.

"All the officers and men on board of the Telegraph Fleet, seem to take the greatest interest in our enterprise, and are very desirous to go out in the ships the next time.

"Since my arrival, I have received the greatest kindness and attention from all whom I have met, from the Lord Lieutenant of Ireland, down to the cabin-boys and sailors. The inclosed letter from the Knight of Kerry, I received with a basket of hothouse fruit, just as we were getting ready to leave Valentia harbor.

Your ———

"CYRUS W. FIELD."

The day that this was written, Mr. Field landed at Portsmouth, and at once hastened to London to meet the Directors. At first it was a question if they should renew the expedition this year. But their brief experience had shown the need of more ample preparations for their next attempt. They required six hundred miles more of cable to make up for over three hundred lost in the sea, and to provide a surplus so as to run no risk of falling short from other accidents; and most of all they needed better machinery to pay out the cable into the ocean. These preparations required time, and before they could be made, it would be late in the autumn. Hence they reluctantly decided to defer the expedition till another year. The Niagara and the Agamemnon therefore discharged their cable at Plymouth, whence the Niagara returned home; and

THE EXPEDITION OF 1857.

Mr. Field, after remaining a few weeks in London to complete the preparations for the next year, sailed for America.

He returned to find that a commercial hurricane had swept over the country, in which a thousand stately fortunes had gone down, and in which the wealth he had accumulated by years of toil had nearly suffered shipwreck. Such were the tidings that met him on landing. It had been a year of disappointments in England and America—of disasters on land and sea— and all his high hopes were

<blockquote>In the deep bosom of the ocean buried.</blockquote>

CHAPTER IX.

THE FIRST EXPEDITION OF 1858.

The expedition of 1857 was little more than an experiment on a grand scale. As such it had its use; but its abrupt ending within three hundred miles of the Irish coast was a severe shock to public confidence. Up to that time the enterprise had been accepted by the people of England and of America, almost without considering its magnitude and difficulty. They had taken it for granted as a thing which must some day be accomplished by human skill and perseverance. But now it had been tried and failed. This first expedition opened their eyes to the vastness of the undertaking, and led many to doubt who did not doubt before. Some even began to look upon it as a romantic adventure of the sea, rather than a serious undertaking. This decline of popular faith was felt as soon as there was a call for more money. Men reasoned that if the former attempt was but an experiment, it was rather a costly one. The loss of three hundred and thirty-five miles of cable, with the postponement of the expedition to another year, was equivalent to a loss of a hundred thousand pounds. To make this good, the

THE FIRST EXPEDITION OF 1858.

Directors had to enlarge the capital of the Company. This new capital was not so readily obtained. Those who had subscribed before, thought they had lost enough; and the public stood aloof till they could see the result of the next experiment. The projectors found that it was easy to go with the current of popular enthusiasm, but very hard to stem a growing popular distrust. They found how great an element of success in all public enterprises is public confidence.

But against this very revulsion of feeling they had been already warned. The Earl of Carlisle the year before had cautioned them against being too sanguine of immediate results, and reminded them that "preliminary failure was even the law and condition of ultimate success." There were many who now remembered his words, and on whom the lesson was not lost.

But whatever the depression at the failure of the first attempt to lay a telegraph across the ocean, and at the thick-coming disasters on land and sea, it did not interfere with renewed and vigorous efforts to prepare for a second expedition. The Directors gave orders for the manufacture of seven hundred miles of new cable, to make up for the loss of the previous year, and to provide a surplus against all contingencies. And the Government promised again its powerful aid.

In America, Mr. Field went to Washington to ask a second time the use of the ships, which had already represented the country so well. He made also a special request for the services of Mr. William E. Everett. This gentleman had been the chief-engineer of the Niagara the year before. He had watched closely the paying-out machine, as it was put together on the deck, and as it worked on the voyage, and with the eye of a practised mechanic, he saw that it required great alterations. It was too cumbrous, had too many wheels, and especially its brakes shut down with a gripe that would snap the strongest chain cable.* Mr. Field saw that this was the man to remedy the defects of the old machine, and to make one that would work more smoothly. He therefore applied especially for his services. To the credit of the administration, it granted both requests in the most handsome manner. "There," said the Secretary of the Navy, handing Mr. Field the official letter, "I have given you all you asked."

After such an answer he did not wait long. The letter is dated the thirtieth of December, and in just one week, on the sixth of January, he sailed in the Persia for England with Mr. Everett. Scarcely had

* It should be said, however, in justice to Mr. Bright, that most of these defects he had himself perceived on seeing it in operation. On his return from the expedition of 1857, he sent in a report, pointing out the defects of the machinery, and how to remedy them. These suggestions were approved by the Scientific Committee, and carried out by Mr. Everett. The recognition of this fact, while it takes nothing from the practical skill shown by the American engineer, is but just to his predecessor, who, as the pioneer in this work, might easily fall into mistakes, which it needed only time and experience to correct.

THE FIRST EXPEDITION OF 1858.

he arrived in London before he was made the General Manager of the Company, with control of the entire staff, including electricians and engineers. The following extract from the minutes of the Board of Directors, dated January 27, 1858, explains the new position to which he was invited:

"The Directors having for several months felt that it would greatly advance the interests of this enterprise, if Mr. Cyrus W. Field, of New York, could be induced to come over to England, for the purpose of undertaking the general management and supervision of all the various arrangements that would be required to be carried out before the sailing of the next expedition; application was made to Mr. Field, with the view of securing his consent to the proposal, and he arrived in this country on the sixteenth instant, when it was ascertained that he would be willing, if unanimously desired by the Directors, to act in behalf of the Company as proposed; and Mr. Field having retired, it was unanimously resolved to tender him, in respect to such services, the sum of £1000 over and above his travelling and other expenses, as remuneration."

This resolution was at once communicated to Mr. Field, who replied that he would undertake the duties of General Manager, but declined the offer of £1000, preferring to give his services to the Company without compensation. Whereupon the Directors immediately passed another resolution:

"That Mr. Field's kind and generous offer be accepted by this Board ; and that their best thanks are hereby ten-dered to him for his devotion to the interests of the undertaking."

The following, passed a few weeks later, March 26, was designed to emphasize the authority given over all the employés of the Company:

"*Resolved*, That Mr. Cyrus W. Field, General Manager of the Company, is hereby authorized and empowered to give such directions and orders to the officers composing the staff of the Company, as he may from time to time deem necessary and expedient with regard to all matters connected with the business proceedings of the Company, subject to the control of the Directors.

"*Resolved*, That the staff of the Company be notified hereof, and required to observe and follow such directions as may be issued by the General Manager."

As Mr. Field was thus invested with the entire charge of the preparations for the next expedition, he was made responsible for it, and felt it due alike to himself and to the Company to omit no means to insure success. It was therefore his duty to examine into every detail. The manufacture of the new cable was already under way, and there was no opportunity to make any change in its construction, even if any had been desired. But there was another matter which was quite as important to success—the construction of the paying-out machines. This had been the great defect of the previous year, and, while it continued, would render success almost impossible. No matter how many hundreds or thousands of miles of cable might be made, if the machinery was not fitted to pay it out into the sea, it would be constantly broken. To remedy these defects was an object of anxious solicitude, and to this the new manager gave his first attention. Hardly was he in London before Mr. Everett was installed at the large machine works of Easton and Amos, in Southwark, where, surrounded by plans and models, he devoted himself for three months to studying out a better invention for this most important work. At the end of that time he had a model complete, and invited a number of the most eminent engineers of London to witness its operation. Among these were Mr. Brunel, and Messrs. Lloyd, Penn, and Field, who had given the enterprise the benefit of their counsel for months, refusing all compensation; Mr. Charles T. Bright, the engineer of the Company, and his two assistants, Mr. Canning and Mr. Clifford, and Mr. Follansbee, the chief-engineer of the Niagara, in the place Mr. Everett had occupied the year before. The machine was set in motion, and all saw its operation, while Mr. Everett explained its parts, and the difficulties which he had tried to overcome. It was obvious at a glance that it was a great improvement on that of the former year. It was much smaller and lighter. It would take up only about one-third of the room on the deck, and had only one-fourth the weight of the old machine. Its construction was much more simple. Instead of four heavy wheels, it had but two, and these were made to revolve with ease, and without danger of sudden check, by the application of what were known as self-releasing brakes. These were the invention of Mr. Appold, of London, a gentleman of fortune, but with a strong taste for mechanics, which led him to spend his time and wealth in exercising his mechanical ingenuity. These brakes were so adjusted as to bear only a certain strain, when they released themselves. This ingenious contrivance was applied by Mr. Everett to the paying-out machinery. The strength of the cable was such that it would not break except under a tension of a little over three tons. The machinery was so adjusted that not more than half that strain could possibly come upon the cable, when the brakes would relax their grasp, the wheels revolve easily, and the cable run out into the sea without a jar. The paying-out machine, therefore, we are far from claiming as wholly an American invention. This part of the mechanism was English. The merit of Mr. Everett lay in the skill with which he adapted it to the laying of the Atlantic cable, and in his improvements of other parts of the machinery. The whole construction, as it afterwards stood upon the decks of the Niagara and the Agamemnon, was the product of English and American invention combined. The engineers, who now saw it for the first time, were delighted. It seemed to have the intelligence of a

THE FIRST EXPEDITION OF 1858. 149

human being, to know when to hold on, and when to let go. All felt that the great difficulty in laying the cable was removed, and that under this gentle manipulation it would glide easily and smoothly from the ship into the sea.

While these preparations were going on in London, the Niagara, which did not leave New York till the ninth of March, arrived at Plymouth, under command of Captain Hudson, to take on board her share of the cable. Both ships had discharged their burden at Keyham Docks, where the precious freight was passed through a composition of tar and pitch and linseed-oil and beeswax, to preserve it from injury, and had been coiled under cover to be kept safely through the winter. The Agamemnon was already at Plymouth, having been designated by the Admiralty again to take part in the work, though under a new commander, Captain George W. Preedy, an excellent officer. The place of the Leopard was taken by the Gorgon, under command of Captain Dayman, who had made the deep-sea soundings in the Cyclops the year before.

While the English Government was thus prompt in furnishing its ships, news arrived from America that the Company could not have again the assistance of the Susquehanna, which had accompanied the Niagara on the preceding expedition. She was in the West-Indies, and the yellow fever had broken out on board. What should be done? It was late to apply again to the American Government, and it was doubtful what would be the result of the application. This threatened some embarrassment. Mr. Field resolved the difficulty in a way which showed his confidence in the great and generous Government on the other side of the water, with which he had occasion so often to deal. Without waiting for the action of the Company, he called a cab, and drove straight to the Admiralty, and sent in his card to Sir John Pakington, then First Lord of the Admiralty. This gentleman, like his predecessor, Sir Charles Wood, had shown the most friendly interest in the Atlantic Telegraph, and given it his warmest support. Mr. Field was received at once, and began with true American eagerness: "I am ashamed to come to you, after what the English Government has done for us. But here is our case. We are disappointed in the Susquehanna. She is in the West-Indies, with the yellow fever on board. She cannot come to England to take part in the expedition. Can you do anything for us?" Sir John replied that the Government had not ships enough for its own use; that it was at that very moment chartering vessels to take troops to Malta—"but he would see what he could do." In an hour or two he sent word to the office of the Company, that Her Majesty's ship Valorous—commanded by Captain W. C. Aldham, an officer of great experience—had been ordered to take the place of the Susquehanna in the next expedition. We mention this

THE FIRST EXPEDITION OF 1858.   151

little incident, not so much to illustrate Mr. Field's prompt and quick manner of deciding and acting, as to show the noble and generous spirit in which the English Government responded to every appeal.

The reshipping of the cable at Plymouth occupied the whole month of April and part of May. Some changes were made in the mode adopted, it being coiled around large cones. The work was done as before, by a hundred and sixty men detailed for the purpose, of whom one fourth were the workmen of the Company, and the rest sailors who had volunteered for the duty. These were divided into gangs of forty, that relieved each other, by which the work went on day and night. In this way they coiled about thirty miles in the twenty-four hours. Owing to the increased length of cable, and the greater care in coiling, it took a longer time than the year before. The whole was completed about the middle of May. There was then in all, on board the two ships, a little over three thousand statute miles. This included, besides seven hundred miles of new cable, thirty-nine miles of that lost the year before, which had been recovered by the Company, and a few miles of condemned cable from Greenwich, which was put on board for experiments. The shipment being thus complete, and the paying-out machines in position, the ships were ready to make a trial trip, preparatory to their final departure.

For this purpose the telegraphic squadron sailed

from Plymouth on Saturday, the twenty-ninth of May, and bore southward two or three hundred miles, till the green color of the sea changing to a deep blue, showed that they had reached the great depths of the ocean. They were now in the waters of the Bay of Biscay, where the soundings were over twenty-five hundred fathoms. Here the Niagara and the Agamemnon were connected by a hawser, being about a quarter of a mile apart. The cable was then passed from one to the other, and a series of experiments begun, designed to test both the strength of the cable and the working of the machinery. Two miles of the cable were paid out, when it parted. This would have seemed a bad sign, had it been any other part of the cable than that which was known to be imperfect and had long since been condemned. The next day three miles were paid out. This, too, was broken, but only when they tried to haul it in, and under a pressure of several tons.

Other experiments were tried, such as splicing the cable, and lowering it to the bottom of the sea—an operation which it was thought might be critical in mid-ocean, but which was performed without difficulty—and running out the cable at a rapid rate, when the speed of the ship was increased to seven knots, without causing the cable to break, or even to kink. On the whole, the result of the trip was satisfactory. The paying-out machine of Mr. Everett worked well, and

THE FIRST EXPEDITION OF 1858. 153

the electric continuity through the whole cable was perfect. It was on this expedition that was used for the first time the marine mirror galvanometer of Professor Thomson, by whom it had been invented for marine testing within the previous four weeks, and which afterwards proved an instrument so important to the success of ocean telegraphy. After these experiments the squadron returned to Plymouth.

As it happened, the present writer had just arrived in England, and landing at Falmouth, hastened to Plymouth, where the ships were lying in the Sound. It was Saturday, the fifth of June, and the next day, by invitation of Captain Hudson, he conducted Divine service on the Niagara, where an awning was spread over the quarter-deck, round which were grouped the officers of the ship, behind whom were crowded four or five hundred seamen. If it was a pleasure in such circumstances to speak to one's countrymen, it was not less to be received with equal kindness on board the Agamemnon. To see these two mighty ships of war, with their consorts, lying side by side, not with guns run out, but engaged in a mission of peace, seemed indeed an omen of the good time coming, when nations shall learn war no more.

Among the matters of personal solicitude and anxiety at this time—next to the success of the expedition—was Mr. Field himself. He was working with an activity which was unnatural—which could only be kept up by great excitement, and which involved the most serious danger. The strain on the man was more than the strain on the cable, and we were in fear that both would break together. Often he had no sleep, except such as he caught flying on the railway. Indeed, when we remonstrated, he said he could rest better there than anywhere else, for then he was not tormented with the thought of any thing undone. For the time being he could do no more; and putting his head in the cushioned corner of the carriage, he got an hour or two of broken sleep.

Of this activity we had an instance while in Plymouth. The ships were then lying in the Sound, only waiting orders from the Admiralty to go to sea; but some business required one of the Directors to go to Paris, and as usual, it fell upon Mr. Field. He left on Sunday night and went to Bristol, and thence, by the first morning train, to London. Monday he was busy all day, and that night went to Paris. Tuesday, another busy day, and that night back to London. Wednesday, occupied every minute till the departure of the Great Western train. That night back to Plymouth. Thursday morning on board the Niagara, and immediately the squadron sailed.

It was the tenth day of June that the expedition left England, with fair skies and bright prospects. In truth, it was a gallant sight, as these four ships stood out to sea together—those old companions, the Ni-

THE FIRST EXPEDITION OF 1858. 155

agara and the Agamemnon, leading the way, followed by their new attendants, the Valorous and the Gorgon. Never did a voyage begin with better omens. The day was one of the mildest of June, and the sea so still, that one could scarcely perceive, by the motion of the ship, when they passed beyond the breakwater off Plymouth harbor into the Channel, or into the open sea. At night, it was almost a dead calm. The second day was like the first. There was scarcely wind enough to swell the sails. The ships were all in sight, and as they kept under easy steam, they seemed bound on a voyage of pleasure, gliding over a summer sea to certain success.

It had been supposed that the expedition of this year would have a great advantage over the last, from sailing two months earlier, at what was considered a more favorable season. So said all the wise men of the sea. They had given their opinion that June was the best month for crossing the Atlantic. Then they were almost sure of fair weather. The first three days of the voyage confirmed these predictions, and they who had made them, being found true prophets, shook their heads with great satisfaction.

But alas! for the vanity of human expectations, or for those who put trust in the treacherous sea. On Sunday it began to blow. The barometer fell, and all signs indicated to the eye of a seaman rough weather. From this time they had a succession of gales for more than a week. From day to day it blew fiercer than before, till Sunday, the twentieth, when the gale was at its height, and the spirit of the storm was out on the Atlantic. Up to this time the Niagara and the Agamemnon (though they had long since parted from the Valorous and the Gorgon) had managed to keep in sight of each other; and now from the deck of the former the latter was seen a mile and a half distant, rolling heavily in the sea. The signals which she made showed that she was struggling with the fury of the gale. She was really in great danger of foundering. But this was owing, not merely to the severity of the gale, but to the enormous weight she carried, and to the way this huge bulk was stowed in the ship. Only a few days before we had been on board of her, and Captain Preedy showed us, in one coil, thirteen hundred miles of cable! This made a dead weight of as many hundred tons, which rendered her in rough weather almost unmanageable. To make the matter worse, she had another coil of about two hundred and fifty tons on the forward deck, where it made the head of the ship heavy. In her tremendous rolls, this coil broke loose, and threatened at a time to dash like an avalanche through the side of the ship. But at the most fearful moments the gallant seaman in command never lost his presence of mind. He was always on deck, watching with a vigilant eye the raging of the tempest, and issuing his orders with

THE FIRST EXPEDITION OF 1858. 157

coolness and prompt decision. To this admirable skill was due the safety of the ship, and of all on board.*

But all things have an end; and this long gale at last blew itself out, and the weary ocean rocked itself to rest. Toward the last of the week the squadron got together at the appointed rendezvous in mid-ocean. As the ships came in sight, the angry sea went down; and on Friday, June twenty-fifth, just fifteen days from Plymouth, they were all together, as tranquil in the middle of the Atlantic as if in Plymouth Sound. "This evening the four vessels lay together, side by side, and there was such a stillness in the sea and air, as would have seemed remarkable in an inland lake; on the Atlantic, and after what we had all so lately witnessed, it seemed almost unnatural." The boats were out, and the officers were passing from ship to ship, telling their experiences of the voyage, and forming their plans for the morrow. Captains Aldham and Dayman said it was the worst weather they had ever experienced in the North-Atlantic. But it was the Agamemnon that suffered most. The rough sea

---

* As there is no trouble without a compensation, it is something that this voyage, fearful as it was, furnished a subject for a description of marvellous power. The letter to the London Times, written by Mr. Woods, its correspondent on board the Agamemnon, is one of the finest descriptions of a storm at sea we know of in the language. It is a wonderful specimen of "word-painting," and brings the scene before us with a vividness like that of the marine paintings of Stanfield or Turner.

had shaken not only the ship, but the cable in her. The upper part of the main coil had shifted, and become so twisted and tangled, that a hundred miles had to be got out and coiled in another part of the ship, so that it was not till the afternoon of Saturday, the twenty-sixth, that the splice was finally made, and the cable lowered to the bottom of the sea. The ships were then got under way, but had not gone three miles, before the cable broke, being caught in the machinery on board the Niagara. It was fortunate they had gone no farther. Both ships at once turned about and spliced again the same afternoon, and made a fresh start. Now all went well. The paying-out machines worked smoothly, and the cable ran off easily into the sea. Thus each ship had paid out about forty miles when suddenly the current ceased!

Says the writer on the Agamemnon: "At half-past three o'clock [Sunday morning] forty miles had gone, and nothing could be more perfect and regular than the working of every thing, when suddenly Professor Thomson came on deck, and reported a total break of continuity; that the cable in fact had parted, and, as was believed at the time, from the Niagara. In another instant a gun and a blue-light warned the Valorous of what had happened, and roused all on board the Agamemnon to a knowledge that the machinery was silent, and that the first part of the Atlantic Cable had been laid and lost effectually."

THE FIRST EXPEDITION OF 1858.

This was disheartening, but not so much from the fact of a fresh breaking of the cable, as from the mystery as to its cause. The fact, of course, was known instantly on both ships, but the cause was unknown. Those on each ship supposed it had occurred on the other. With this impression they turned about to beat up again toward the rendezvous. It was noon of Monday, the twenty-eighth, before the Agamemnon rejoined the Niagara; and then, says the writer:

"While all were waiting with impatience for her explanation of how they broke the cable, she electrified every one by running up the interrogatory : 'How did the cable part ?' This was astounding. As soon as the boats could be lowered, Mr. Cyrus Field, with the electricians from the Niagara, came on board, and a comparison of logs showed the painful and mysterious fact that, *at the same second of time,* each vessel discovered that a total fracture had taken place at a distance of certainly not less than ten miles from each ship; in fact, as well as can be judged, at the bottom of the ocean. That of all the many mishaps connected with the Atlantic Telegraph, this is the worst and most disheartening is certain, since it proves that, after all that human skill and science can effect to lay the wire down with safety has been accomplished, there may be some fatal obstacles to success at the bottom of the ocean, which can never be guarded against; for even the nature of the peril must always remain as secret and unknown as the depths in which it is to be encountered."

But it was no time for useless regrets. Once more the cable was joined in mid-ocean, and dropped to its silent bed, and the Niagara and the Agamemnon began to steam away toward opposite shores of the Atlantic. This time the experiment succeeded better than before. The progress of the English ship is thus reported :

"At first, the ship's speed was only two knots, the cable going three and three and a half, with a strain of fifteen hundred pounds. By and by, however, the speed was increased to four knots, the cable going five, at a strain of two thousand pounds. At this rate it was kept, with trifling variations, throughout almost the whole of Monday night, and neither Mr. Bright, Mr. Canning, nor Mr. Clifford ever quitted the machines for an instant. Toward the middle of the night, while the rate of the ship continued the same, the speed at which the cable paid out slackened nearly a knot an hour, while the dynamometer indicated as low as thirteen hundred pounds. This change could only be accounted for on the supposition that the water had shallowed to a considerable extent, and that the vessel was, in fact, passing over some submarine Ben Nevis or Skiddaw. After an interval of about an hour, the strain and rate of progress of the cable again increased, while the increase of the vertical angle seemed to indicate that the wire was sinking down the side of a declivity. Beyond this, there was no variation throughout Monday night, or, indeed, through Tuesday."

On board the Niagara was the same scene of anxious watching every hour of the day and night. Engineers and electricians were constantly on duty :

"The scene at night was beautiful. Scarcely a word was spoken ; silence was commanded, and no conversation

THE FIRST EXPEDITION OF 1858.

allowed. Nothing was heard but the strange rattling of the machine as the cable was running out. The lights about deck and in the quarter-deck circle added to the singularity of the spectacle; and those who were on board the ship describe the state of anxious suspense in which all were held as exceedingly impressive."

Warned by repeated failures, they hardly dared to hope for success in this last experiment. And yet the spirits of all rose, as the distance widened between the ships. A hundred miles were laid safely—a hundred and fifty—two hundred! Why might they not lay two thousand? So reasoned the sanguine and hopeful when, Tuesday night, came the fatal announcement that the electric current had ceased to flow. It afterward appeared that the cable had broken about twenty feet from the stern of the Agamemnon.

As the cable was now useless, it only remained to cut it from the stern of the Niagara. Before doing this, it was thought a good opportunity to test its strength. For this purpose the brakes were shut down, so that the paying-out machine could not move. But still the cable did not break, although the whole weight of the Niagara hung upon that slender cord, and though several men got upon the brakes. Says Captain Hudson: "Although the wind was quite fresh, the cable held the ship for one hour and forty minutes before breaking, and notwithstanding a strain of four tons."

Though not unexpected, this last breaking of the cable was a sad blow to all on board. It was the end of their hopes, at least for the present expedition. Before separating, it had been agreed, that if the cable should part again before either ship had run a hundred miles, they should return and renew the attempt. If they had passed that limit, they were all to sail for Ireland. But the Niagara had run out a hundred and eleven miles, and knowing that the Agamemnon had done about the same, she expected the latter would keep on her course eastward, not stopping till she reached Queenstown. The Niagara, therefore, reluctantly bore away for the same port.

Of course, the return voyage was "any thing but gay." When soldiers come home from the war, they march with a proud step, if they have had a victorious campaign. But it is otherwise when they come with a sad tale of disaster and defeat. Seldom had an expedition begun with higher hopes, or ended in more complete failure. Who could help feeling keenly this fresh disappointment? Even with all the courage "that may become a man," heightened by a natural buoyancy of spirits, how was it possible to resist the impression of the facts they had just witnessed? If—as Lord Carlisle had told them the year before—"there was almost enough of glory in the very design of an Atlantic telegraph," that glory might still be theirs. But apparently they could hope for nothing more.

THE FIRST EXPEDITION OF 1858.

They had done all that men could do. But fate seemed against them; and who can fight against destiny? No one can blame them if they sometimes had sore misgivings, and looked out sadly upon the sea that had baffled their utmost skill, and now laughed their efforts to scorn.

In this mood they entered once more the harbor of Queenstown. The Niagara was the first to arrive and to bring tidings of the great disaster. The Agamemnon came in a few days after. Knowing the fatal impression their report was likely to produce, Mr. Field hastened to London to meet the Directors. It was high time. The news had reached there before him, and had already produced its effect. Under its impression the Board was called together. It met in the same room where, six weeks before, it had discussed the prospects of the expedition with full confidence of success. Now it met, as a council of war is summoned after a terrible defeat, to decide whether to surrender or to try once more the chances of battle. When the Directors came together, the feeling—to call it by the mildest name—was one of extreme discouragement. They looked blankly in each other's faces. With some, the feeling was one almost of despair. Sir William Brown, of Liverpool, the first Chairman, wrote, advising them to sell the cable. Mr. Brooking, the Vice-Chairman, who had given more time to it than any other Director, when he saw that his colleagues were disposed to make still another trial, left the room, and the next day sent in his resignation, determined to take no further part in an undertaking which had been proved hopeless, and to persist in which seemed mere rashness and folly.

But others thought there was still a chance. Like Robert Bruce, who, after twelve battles and twelve defeats, yet believed that a thirteenth *might* bring victory, they clung to this bare possibility. Mr. Field and Professor Thomson gave the results of their experience, from which it appeared that there was no obstacle in the nature of the case which might not be overcome. Mr. Bright and Mr. Woodhouse joined with them in advising strongly that they should renew the attempt. To be sure, it was a forlorn hope. But the ships were there. Though they had lost three hundred miles of cable, they had still enough on board to cross the sea. These arguments prevailed, and it was voted to make one more trial before the project was finally abandoned. Even though the chances were a hundred to one against them, that one might bring them success. And so it proved. But was it their own wisdom or courage that got them the victory, or were they led by that Being whose way is in the sea, and whose path is in the great waters?

CHAPTER X.

THE SECOND EXPEDITION SUCCESSFUL.

A BOLD decision needs to be followed by prompt action, lest the spirit that inspired it be weakened by delay. When once it had been fixed that there was to be another attempt to lay the Atlantic cable, no time was lost in carrying the resolve into execution. The telegraphic fleet was lying at Queenstown. The Niagara had arrived on the fifth of July, but the Agamemnon, which, through some misunderstanding, had returned to the rendezvous in mid-ocean, thus crossing the Niagara on her track, did not get in till a week later. However, all were now there, safe and sound, and Mr. Field and Mr. Samuel Gurney went to the Admiralty, and got an order which they telegraphed to the ships to get ready immediately to go to sea. Not an hour was lost. They had barely time to take in coal and other supplies for the voyage. Mr. Field hastened from England, and Prof. Thomson from his home in Scotland, and in five days the squadron was under way, bound once more for the middle of the Atlantic.

It was Saturday, the seventeenth of July, that the

ships left on their second expedition. As they sailed out of the Cove of Cork, it was with none of the enthusiasm which attended their departure from Valentia the year before, or even from Plymouth on the tenth of June. Nobody cheered; nobody bade them God-speed. "As the ships left the harbor, there was apparently no notice taken of their departure by those on shore, or in the vessels anchored around them; every one seemed impressed with the conviction that they were engaged in a hopeless enterprise, and the squadron seemed rather to have slunk away on some discreditable mission, than to have sailed for the accomplishment of a grand national scheme." Many even of those on board felt that they were going on a fool's errand; that the Company was possessed by a kind of insanity, of which they would soon be cured by another bitter experience.

On leaving this second time, it was agreed that the squadron should not try to keep together, but each ship make its way to the given latitude and longitude which was the appointed rendezvous in mid-ocean. The Niagara, being the largest, and able to carry the most coal, kept under steam the whole way, and arrived first, and waited several days for the other ships to appear. The Valorous came next, and then the Gorgon, and, last of all, the Agamemnon, which had been saving her coal for the return voyage, and had been delayed for want of a little of that wind which,

THE SECOND EXPEDITION SUCCESSFUL. 167

in the former expedition, she had had in too great abundance. Says the English correspondent on board :

"For several days in succession there was an uninterrupted calm. The moon was just at the full, and for several nights it shone with a brilliancy which turned the sea into one silvery sheet, which brought out the dark hull and white sails of the ship in strong contrast to the sea and sky, as the vessel lay all but motionless on the water, the very impersonation of solitude and repose. Indeed, until the rendezvous was gained, we had such a succession of beautiful sunrises, gorgeous sunsets, and tranquil moonlight nights, as would have excited the most enthusiastic admiration of any one but persons situated as we were. But by us such scenes were regarded only as the annoying indications of the calm, which delayed our progress and wasted our coal. By dint, however, of a judicious expenditure of fuel, and a liberal use of the cheaper motive power of sail, the rendezvous was reached on Wednesday, the twenty-eighth of July, just eleven days after our departure from Queenstown. The rest of the squadron came in sight at nightfall, but at such a distance that it was past ten o'clock on the morning of Thursday, the twenty-ninth, before the Agamemnon joined them.

"The day was beautifully calm, so no time was to be lost before making the splice ; boats were soon lowered from the attendant ships, the two vessels made fast by a hawser, and the Niagara's end of the cable conveyed on board the Agamemnon. About half-past twelve o'clock the splice was effectually made. In hoisting it out from the side of the ship the leaden sinker broke short off and fell overboard ; and there being no more convenient weight at hand, a thirty-two

168 STORY OF THE ATLANTIC TELEGRAPH.

pound shot was fastened to the splice instead, and the whole apparatus was quickly dropped into the sea without any formality, and indeed almost without a spectator, for those on board the ship had witnessed so many beginnings to the telegraphic line, that it was evident they despaired of there ever being an end to it. The stipulated two hundred and ten fathoms having been paid out, the signal to start was hoisted, the hawser cast loose, and the Niagara and Agamemnon started for the last time for their opposite destinations."

At this moment the ships were nearly in mid-ocean, but not exactly. Mr. Field, who never indulged in poetical descriptions, but always gave the figures, stating the precise latitude and longitude, and from what quarter the wind blew, and how many fathoms deep the ocean was, and how many miles of cable were on board, made the following entry in his journal:

"Thursday, July twenty-ninth, latitude fifty-two degrees nine minutes north, longitude thirty-two degrees twenty-seven minutes west. Telegraph Fleet all in sight; sea smooth; light wind from S.E. to S.S.E.; cloudy. Splice made at one P.M. Signals through the whole length of the cable on board both ships perfect. Depth of water fifteen hundred fathoms; distance to the entrance of Valentia harbor eight hundred and thirteen nautical miles, and from there to the telegraph-house the shore end of the cable is laid. Distance to the entrance of Trinity Bay, Newfoundland, eight hundred and twenty-two nautical miles, and from there to the telegraph-house at the head of the bay of Bull's

THE SECOND EXPEDITION SUCCESSFUL. 169

Arm, sixty miles, making in all eight hundred and eighty-two nautical miles. The Niagara has sixty-nine miles further to run than the Agamemnon. The Niagara and Agamemnon have each eleven hundred nautical miles of cable on board, about the same quantity as last year."

And now, as the ships are fairly apart, and will soon lose sight of each other, we will leave the Agamemnon for the present to pursue her course toward Ireland, while we follow our own Niagara to the shores of the New World. At first of course, while all hoped for success, no one dared to expect it. They said afterwards that "Mr. Field was the only man on board who kept up his courage through it all." But the chances seemed many to one against them; and the warnings were frequent to excite their fears. That very evening, about sunset, all again seemed lost. We quote from Mr. Field's journal: " At forty-five minutes past seven P.M., ship's time, signals from the Agamemnon ceased, and the tests applied by the electricians showed that there was a want of continuity in the cable, but the insulation was perfect. Kept on paying out from the Niagara very slowly, and constantly applying all kinds of electrical tests until ten minutes past nine, ship's time, when again commenced receiving perfect signals from the Agamemnon." At the same moment the English ship had the same relief from anxiety.

The next day there was a fresh cause of alarm. It

was found that the Niagara had run some miles out of her course. Comparing the distance run by observation and by patent log, there was a difference of sixteen miles and a third. With such a percentage of loss, the cable would not hold out to reach Newfoundland. This was alarming, but it was soon explained. The mass of iron in the ship had affected the compass, so that it no longer pointed to the right quarter of the heavens. Had the Niagara been alone on the ocean, this might have caused serious trouble. But now appeared the great advantage of an attendant ship. It was at once arranged that the Gorgon should go ahead and lead the way. As she had no cable on board, her compasses were subject to no deviation. Accordingly she took her position in the advance, keeping the line along the great circle arc, which was the prescribed route. From that moment there was no variation, or but a very slight one. The two methods of computing the distance—by log and by observation—nearly coincided, and the ship varied scarcely a mile from her course till she entered Trinity Bay.

It is not necessary to follow the whole voyage, for the record is the same from day to day. It is the same sleepless watching of the cable as it runs out day and night, and the same anxious estimate of the distance that still separates them from land. Communication is kept up constantly between the ships. Mr. Field's journal contains entries like these:

SECOND EXPEDITION SUCCESSFUL.

"Saturday, July thirty-first. By eleven o'clock had paid out from the Niagara three hundred miles of cable; at forty-five minutes past two received signals from the Agamemnon that they had paid out from her three hundred miles of cable; at thirty-seven minutes past five finished coil on the berth-deck, and commenced paying out from the lower deck."

"Monday, August second. The Niagara getting light, and rolling very much; it was not considered safe to carry sail to steady ship, for in case of accident it might be necessary to stop the vessel as soon as possible. Passed and signalled the Cunard steamer from Boston to Liverpool." Same day about noon, "imperfect insulation of cable detected in sending and receiving signals from the Agamemnon, which continued until forty minutes past five, when all was right again. The fault was found to be in the ward-room, about sixty miles from the lower end, which was immediately cut out, and taken out of the circuit."

"Tuesday, August third. At a quarter-past eleven, ship's time, received signals from on board the Agamemnon, that they had paid out from her seven hundred and eighty miles of cable. In the afternoon and evening passed several icebergs. At ten minutes past nine P.M., ship's time, received signal from the Agamemnon that she was in water of two hundred fathoms. At twenty minutes past ten P.M., ship's time, Niagara in water of two hundred fathoms, and informed the Agamemnon of the same.

"Wednesday, August fourth. Depth of water less than two hundred fathoms. Weather beautiful, perfectly calm. Gorgon in sight. Sixty-four miles from the telegraph-house. Received signal from Agamemnon at noon that they had paid out from her nine hundred and forty miles of cable.

Passed this morning several icebergs. Made the land off entrance to Trinity Bay at eight A.M. Entered Trinity Bay at half-past twelve. At half-past two, we stopped sending signals to Agamemnon for fourteen minutes, for the purpose of making splice. At five P.M. saw Her Majesty's steamer Porcupine [which had been sent by the British Government to Newfoundland, to watch for the telegraph ships] coming to us. At half-past seven, Captain Otter, of the Porcupine, came on board of the Niagara to pilot us to the anchorage, near the telegraph-house.*

* The spot chosen as the terminus of the Atlantic cable, with the views around it—both on the water and on land—is thus described by a correspondent:

"All who have visited Trinity Bay, Newfoundland, with one consent allow it to be one of the most beautiful sheets of water they ever set eyes upon. Its color is very peculiar—an inexpressible mingling of the pure blue ocean with the deep evergreen woodlands and the serene blue sky. Its extreme length is about eighty miles, its breadth about thirty miles, opening boldly into the Atlantic on the northern side of the island. At its south-western shore it branches into the Bay of Bull's Arm, which is a quiet, safe, and beautiful harbor, about two miles in breadth, and nine or ten in length, running in a direction north-west.

"The depth of water is sufficient for the largest vessels. The tide rises seven or eight feet, and the bay terminates in a beautiful sand-beach. The shore is clothed with dark green fir-trees, which, mixed with birch and mountain-ash, present a pleasing contrast. The land rises gradually from the water all around, so as to afford one of the most agreeable town sites in the island. You ascend only about a quarter of a mile from the water, and there are no longer trees, but wild grass like an open prairie. Here are found at this season myriads of the upland cranberries, upon which unnumbered ptarmigan, or the northern partridge, feed.

"The raspberry, bake-apple berry, and the whortleberry are also common. Numerous little lakes may be seen in the open, elevated ground, from which flow rivulets, affording abundance of fine trout.

SECOND EXPEDITION SUCCESSFUL.

"Thursday, August fifth. At forty-five minutes past one A.M., Niagara anchored. Total amount of cable paid out since splice was made, ten hundred and sixteen miles, six hundred fathoms. Total amount of distance, eight hundred and eighty-two miles. Amount of cable paid out over distance run, one hundred and thirty-four miles, six hundred fathoms, being a surplus of about fifteen per cent. At two A.M., I went ashore in a small boat, and awoke persons in charge of the telegraph-house, half a mile from landing, and informed them that the Telegraph Fleet had arrived, and were ready to land the end of the cable. At forty-five minutes past two, received signal from the Agamemnon that she had paid out ten hundred and ten miles of cable. At

After ascending for about a mile and a half, you are then probably three hundred or four hundred feet above the tide, and nothing can exceed the beauty of the scene when, at one view, you behold the placid waters of both Trinity and Placentia Bays—the latter sprinkled with clusters of verdant islands.

"You can now descend westward as gradually as you came up from the Telegraph landing, to the shores of Placentia Bay, where there is an excellent harbor and admirable fisheries, skirting the shore, and the accompanying road of the land Telegraph line leading from St. John's westward through the island, to Cape Ray. At this season of the year game is very abundant. Reindeer in great numbers, bears, wolves—others very numerous, the large northern hare, foxes, wild geese, ducks, etc.

"About four miles southward of the entrance of the bay of Bull's Arm, on the shore of Placentia Bay, is situated the extraordinary La Manche lead mine, the property of the Telegraph Company, already yielding a rich supply of remarkably pure galena. The place where the cable is landed is memorable in the history of the island as the naval battle-ground between the French and English in their early struggle for the exclusive occupancy of the valuable fisheries along the coast."

four A.M., delivered telegraphic despatch for the Associated Press, to be forwarded to New York as early in the morning as the offices of the line were open.

"At a quarter-past five A.M., telegraph cable landed. At six, end of cable carried into telegraph-house, and received very strong currents of electricity through the whole cable from the other side of the Atlantic. Captain Hudson, of the Niagara, then read prayers, and made some remarks.

"At one P.M., Her Majesty's steamer Gorgon fired a royal salute of twenty-one guns."

Thus simply was the story told, that in a few hours was to send a thrill throughout the continent.

To complete the narrative of the expedition, it is necessary to include the voyage of the Agamemnon, the best account of which is given in the letter of the correspondent of the London Times. We quote from the time of junction in mid ocean, just as the ships went sailing eastward and westward:

"For the first three hours the ships proceeded very slowly, paying out a great quantity of slack, but after the expiration of this time, the speed of the Agamemnon was increased to about five knots per hour, the cable going at about six, without indicating more than a few hundred pounds of strain upon the dynamometer. Shortly after six o'clock a very large whale was seen approaching the starboard bow at a great speed, rolling and tossing the sea into foam all around, and for the first time we felt the possibility of the supposition that our second mysterious breakage of the cable might have been caused after all by one of these animals getting foul of it under water. It appeared as if it

SECOND EXPEDITION SUCCESSFUL.   175

were making direct for the cable, and great was the relief of all when the ponderous living mass was seen slowly to pass astern, just grazing the cable where it entered the water, but fortunately without doing any mischief.

"All seemed to go well up to about eight o'clock; the cable paid out from the hold with an evenness and regularity which showed how carefully and perfectly it had been coiled away; and to guard against accidents which might arise in consequence of the cable having suffered injury during the storm, the indicated strain upon the dynamometer was never allowed to go beyond seventeen hundred pounds, or less than one quarter what the cable is estimated to bear, and thus far every thing looked promising of success. But, in such a hazardous work, no one knows what a few minutes may bring forth, for soon after eight, an injured portion of the cable was discovered about a mile or two from the portion paying out. Not a moment was lost by Mr. Canning, the engineer on duty, in setting men to work to cobble up the injury as well as time would permit, for the cable was going out at such a rate that the damaged portion would be paid overboard in less than twenty minutes, and former experience had shown us that to check either the speed of the ship, or the cable, would, in all probability, be attended by the most fatal results.

"Just before the lapping was finished, Professor Thomson reported that the electrical continuity of the wire had ceased, but that the insulation was still perfect; attention was naturally directed to the injured piece as the probable source of the stoppage, and not a ·moment was lost in cutting the cable at that point, with the intention of making a perfect splice. To the consternation of all, the electrical tests applied showed the fault to be overboard, and in all probability some fifty miles from the ship. Not a second was to be lost, for it was evident that the cut portion must be paid overboard in a few minutes, and in the mean time, the tedious and difficult operation of making a splice had to be performed. The ship was immediately stopped, and no more cable paid out than was absolutely necessary to prevent it breaking.

"As the stern of the ship was lifted by the waves, a scene of the most intense excitement followed. It seemed impossible, even by using the greatest possible speed, and paying out the least possible amount of cable, that the junction could be finished before the part was taken out of the hands of the workmen. The main hold presented an extraordinary scene; nearly all the officers of the ship and of those connected with the expedition, stood in groups about the coil, watching with intense anxiety the cable, as it slowly unwound itself nearer and nearer the joint, while the workmen, directed by Mr. Canning, under whose superintendence the cable was originally manufactured, worked at the splice as only men could work who felt that the life and death of the expedition depended upon their rapidity. But all their speed was to no purpose, as the cable was unwinding within a hundred fathoms, and, as a last and desperate resource, the cable was stopped altogether, and, for a few minutes, the ship hung on by the end. Fortunately, however, it was only for a few minutes, as the strain was continually rising above two tons, and it would not hold on much longer; when the splice was finished, the signal was made to loose the stopper, and it passed overboard safely enough.

"When the excitement consequent upon having so narrowly saved the cable had passed away, we awoke to the

SECOND EXPEDITION SUCCESSFUL. 177

consciousness that the case was still as hopeless as ever, for the electrical continuity was still entirely wanting. Preparations were consequently made to pay out as little rope as possible, and to hold on for six hours, in the hopes that the fault, whatever it might be, might mend itself before cutting the cable and returning to the rendezvous to make another splice. The magnetic needles on the receiving instruments were watched closely for the returning signals; when, in a few minutes, the last hope was extinguished by their suddenly indicating dead earth, which tended to show that the cable had broken from the Niagara, or that the insulation had been completely destroyed.

"In three minutes, however, every one was agreeably surprised by the intelligence that the stoppage had disappeared, and that the signals had again appeared at their regular intervals from the Niagara. It is needless to say what a load of anxiety this news removed from the minds of every one; but the general confidence in the ultimate success of the operations was much shaken by the occurrence, for all felt that every minute a similar accident might occur. For some time the paying-out continued as usual, but toward the morning another damaged place was discovered in the cable; there was fortunately, however, time to repair it in the hold without in any way interfering with the operations beyond for a time slightly reducing the speed of the ship.

"During the morning of Friday, the thirtieth, every thing went well; the ship had been kept at the speed of about five knots, the cable paid out at about six, the average angle with the horizon at which it left the ship being about fifteen degrees, while the indicated strain upon the dynamometer seldom showed more than sixteen hundred pounds to seventeen hundred pounds. Observations made at noon
12

showed that we had made good ninety miles from the starting-point since the previous day, with an expenditure, including the loss in lowering the splice and during the subsequent stoppages, of one hundred and thirty-five miles of the cable. During the latter portion of the day the barometer fell considerably, and toward the evening it blew almost a gale of wind from the eastward, dead ahead of course. As the breeze freshened, the speed of the engines was gradually increased, but the wind more than increased in proportion, so that, before the sun went down, the Agamemnon was going full steam against the wind, only making a speed of about four knots an hour. During the evening topmasts were lowered, and spars, yards, sails, and indeed every thing aloft that could offer resistance to the wind, was sent down on deck; but still the ship made but little way, chiefly in consequence of the heavy sea, though the enormous quantity of fuel consumed showed us that, if the wind lasted, we should be reduced to burning the masts, spars, and even the decks, to bring the ship into Valentia.

"It seemed to be our particular ill-fortune to meet with head-winds whichever way the ship's head was turned. On our journey out we had been delayed, and obliged to consume an undue proportion of coal, for want of an easterly wind, and now all our fuel was wanted because of one. However, during the next day the wind gradually went around to the south-west, which, though it raised a very heavy sea, allowed us to husband our small remaining store of fuel.

"At noon on Saturday, the thirty-first of July, observations showed us to have made good one hundred and twenty miles of distance since noon of the previous day, with a loss of about twenty-seven per cent of cable. The Niagara, as

SECOND EXPEDITION SUCCESSFUL.

far as could be judged from the amount of cable she paid out, which, by a previous arrangement, was signalled at every ten miles, kept pace with us, within one or two miles, the whole distance across. During the afternoon of Saturday, the wind again freshened up, and before nightfall it again blew nearly a gale of wind, and a tremendous sea ran before it from the south-west, which made the Agamemnon pitch to such an extent that it was thought impossible the cable could hold on through the night; indeed, had it not been for the constant care and watchfulness exercised by Mr. Bright, and the two energetic engineers, Mr. Canning and Mr. Clifford, who acted with him, it could not have been done at all. Men were kept at the wheels of the machine to prevent their stopping as the stern of the ship rose and fell with the sea, for, had they done so, the cable must undoubtedly have parted.

"During Sunday the sea and wind increased, and before the evening it blew a smart gale. Now, indeed, were the energy and activity of all engaged in the operation tasked to the utmost. Mr. Hoar and Mr. Moore, the two engineers who had the charge of the relieving-wheels of the dynamometer, had to keep watch and watch alternately every four hours, and while on duty durst not let their attention be removed from their occupation for one moment, for on their releasing the brakes every time the stern of the ship fell into the trough of the sea entirely depended the safety of the cable, and the result shows how ably they discharged their duty. Throughout the night, there were few who had the least expectation of the cable holding on till morning, and many remained awake listening for the sound that all most dreaded to hear—namely, the gun which should announce the failure of all our hopes. But still the cable, which, in comparison with the ship from which it was paid out, and the gigantic waves among which it was delivered, was but a mere thread, continued to hold on, only leaving a silvery phosphorous line upon the stupendous seas as they rolled on toward the ship.

"With Sunday morning came no improvement in the weather; still the sky remained black and stormy to windward, and the constant violent squalls of wind and rain which prevailed during the whole day served to keep up, if not to augment, the height of the waves. But the cable had gone through so much during the night, that our confidence in its continuing to hold was much restored.

"At noon, observations showed us to have made good one hundred and thirty miles from noon of the previous day, and about three hundred and sixty from our starting-point in mid-ocean. We had passed by the deepest sounding of twenty-four hundred fathoms, and over more than half of the deep water generally, while the amount of cable still remaining in the ship was more than sufficient to carry us to the Irish coast, even supposing the continuance of the bad weather should oblige us to pay out the same amount of slack cable we had been hitherto wasting. Thus far things looked very promising for our ultimate success. But former experience showed us only too plainly that we could never suppose that some accident might not arise until the ends had been fairly landed on the opposite shores.

"During Sunday night and Monday morning the weather continued as boisterous as ever, and it was only by the most indefatigable exertions of the engineer upon duty that the wheels could be prevented from stopping altogether, as the vessel rose and fell with the sea, and once or twice they did come completely to a standstill, in spite of all that could be

SECOND EXPEDITION SUCCESSFUL. 181

done to keep them moving ; but fortunately they were again set in motion before the stern of the ship was thrown up by the succeeding wave. No strain could be placed upon the cable, of course ; and though the dynamometer occcasionally registered seventeen hundred pounds as the ship lifted, it was oftener below one thousand, and was frequently nothing, the cable running out as fast as its own weight and the speed of the ship could draw it. But even with all these forces acting unresistedly upon it, the cable never paid itself out at a greater speed than eight knots an hour at the time the ship was going at the rate of six knots and a half. Subsequently, however, when the speed of the ship even exceeded six knots and a half, the cable never ran out so quick. The average speed maintained by the ship up to this time, and, indeed, for the whole voyage, was about five knots and a half, the cable, with occasional exceptions, running about thirty per cent faster.

"At noon on Monday, August second, had made good one hundred and twenty-seven and a half miles since noon of the previous day, and completed more than the half way to our ultimate destination.

"During the afternoon an American three-masted schooner, which afterward proved to be the Chieftain, was seen standing from the eastward toward us. No notice was taken of her at first, but when she was within about half a mile of the Agamemnon she altered her course, and bore right down across our bows. A collision, which might prove fatal to the cable, now seemed inevitable, or could only be avoided by the equally hazardous expedient of altering the Agamemnon's course. The Valorous steamed ahead, and fired a gun for her to heave to, which, as she did not appear to take much notice of, was quickly followed by another from the bows of the Agamemnon, and a second and third from the Valorous, but still the vessel held on her course ; and as the only resource left to avoid a collision, the course of the Agamemnon was altered just in time to pass within a few yards of her. It was evident that our proceedings were a source of the greatest possible astonishment to them, for all her crew crowded upon her deck and rigging. At length they evidently discovered who we were, and what we were doing, for the crew manned the rigging, and dipping the ensign several times they gave us three hearty cheers. Though the Agamemnon was obliged to acknowledge these congratulations in due form, the feelings of annoyance with which we regarded the vessel which, either by the stupidity or carelessness of those on board, was so near adding a fatal and unexpected mishap to the long chapter of accidents which had already been encountered, may easily be imagined. To those below, who of course did not see the ship approaching, the sound of the first gun came like a thunderbolt, for all took it as the signal of the breaking of the cable. The dinner-tables were deserted in a moment, and a general rush made up the hatches to the deck ; but before reaching it, their fears were quickly banished by the report of the succeeding gun, which all knew well could only be caused by a ship in our way or a man overboard.

"Throughout the greater portion of Monday morning the electrical signals from the Niagara had been getting gradually weaker, until they ceased altogether for nearly three-quarters of an hour. Our uneasiness, however, was in some degree lessened by the fact that the stoppage appeared to be a want of continuity,* and not any defect in insulation,

* This is an error, as we learn on the high authority of Professor Thomson himself. It *was* defective insulation, not any " want of con-

and there was consequently every reason to suppose that it might arise from faulty connection on board the Niagara. Accordingly Professor Thomson sent a message to the effect that the signals were too weak to be read, and, as if they had been awaiting such a signal to increase their battery power, the deflections immediately returned even stronger than they had ever been before. Toward the evening, however, they again declined in force for a short time. With the exception of these little stoppages, the electrical condition of the submerged wire seemed to be much improved. It was evident that the low temperature of the water at the immense depth improved considerably the insulating properties of the gutta-percha, while the enormous pressure to which it must have been subjected probably tended to consolidate its texture, and to fill up any air-bubbles or slight faults in manufacture which may have existed.

"The weather during Monday night moderated a little, but still there was a very heavy sea on, which endangered the wire every second minute.

"About three o'clock on Tuesday morning, all on board were startled from their beds by the loud booming of a gun. Every one, without waiting for the performance of the most particular toilet, rushed on deck to ascertain the cause of the disturbance. Contrary to all expectation, the cable was safe, but just in the gray light could be seen the Valorous rounded to in the most warlike attitude, firing gun after gun in quick succession toward a large American bark, which, quite unconscious of our proceeding, was standing right across our stern. Such loud and repeated remonstrances from a large

tinuity," that caused the weak signals. Want of continuity would have stopped the signals altogether, and given quite different indications on the testing instruments from those he observed.

steam frigate were not to be despised, and, evidently without knowing the why or the wherefore, she quickly threw her sails aback and remained hove to. Whether those on board her considered that we were engaged in some filibustering expedition, or regarded our proceedings as another British outrage upon the American flag, it is impossible to say ; but certain it is that, apparently in great trepidation, she remained hove to until we had lost sight of her in the distance.

"Tuesday was a much finer day than any we had experienced for nearly a week, but still there was a considerable sea running, and our dangers were far from passed ; yet the hopes of our ultimate success ran high. We had accomplished nearly the whole of the deep-sea portion of the route in safety, and that, too, under the most unfavorable circumstances possible ; therefore there was every reason to believe that unless some unforeseen accident should occur, we should accomplish the remainder.

"About five o'clock in the evening, the steep submarine mountain which divides the telegraphic plateau from the Irish coast was reached, and the sudden shallowing of the water had a very marked effect upon the cable, causing the strain on and the speed of it to lessen every minute. A great deal of slack was paid out to allow for any great inequalities which might exist, though undiscovered by the sounding-line. About ten o'clock the shoal water of two hundred and fifty fathoms was reached ; the only remaining anxiety now was the changing from the lower main coil to that upon the upper deck, and this most difficult and dangerous operation was successfully performed between three and four o'clock on Wednesday morning.

"Wednesday was a beautiful, calm day ; indeed, it was

SECOND EXPEDITION SUCCESSFUL. 185

the first on which any one would have thought of making a splice since the day we started from the rendezvous. We therefore congratulated ourselves on having saved a week by commencing operations on the Thursday previous. At noon, we were eighty-nine miles distant from the telegraph station at Valentia. The water was shallow, so that there was no difficulty in paying out the wire almost without any loss of slack, and all looked upon the undertaking as virtually accomplished.

"At about one o'clock in the evening, the second change from the upper-deck coil to that upon the orlop-deck was safely effected, and shortly after the vessels exchanged signals that they were in two hundred fathoms water. As the night advanced the speed of the ship was reduced, as it was known that we were only a short distance from the land, and there would be no advantage in making it before daylight in the morning. About twelve o'clock, however, the Skelligs Light was seen in the distance, and the Valorous steamed on ahead to lead us in to the coast, firing rockets at intervals to direct us, which were answered by us from the Agamemnon, though, according to Mr. Moriarty, the master's wish, the ship, disregarding the Valorous, kept her own course, which proved to be the right one in the end.

"By daylight on the morning of Thursday, the bold and rocky mountains which entirely surround the wild and picturesque neighborhood of Valentia, rose right before us at a few miles' distance. Never, probably, was the sight of land more welcome, as it brought to a successful termination one of the greatest, but, at the same time, most difficult schemes which was ever undertaken. Had it been the dullest and most melancholy swamp on the face of the earth that lay before us, we should have found it a pleasant prospect ; but, as the sun rose from the estuary of Dingle Bay, tinging with a deep, soft purple the lofty summits of the steep mountains which surround its shores, and illuminating the masses of morning vapor which hung upon them, it was a scene which might vie in beauty with any thing that could be produced by the most florid imagination of an artist.

"No one on shore was apparently conscious of our approach, so the Valorous steamed ahead to the mouth of the harbor and fired a gun. Both ships made straight for Doulus Bay, and about six o'clock came to anchor at the side of Beginish Island, opposite to Valentia. As soon as the inhabitants became aware of our approach, there was a general desertion of the place, and hundreds of boats crowded around us, their passengers in the greatest state of excitement to hear all about our voyage. The Knight of Kerry was absent in Dingle, but a messenger was immediately dispatched for him, and he soon arrived in Her Majesty's gunboat Shamrock. Soon after our arrival, a signal was received from the Niagara that they were preparing to land, having paid out one thousand and thirty nautical miles of cable, while the Agamemnon had accomplished her portion of the distance with an expenditure of one thousand and twenty miles,* making the total length of the wire submerged two thousand and fifty geographical miles. Immediately after the ships cast anchor, the paddle-box boats of the Valorous were got ready, and two miles of cable coiled away in them, for the purpose of landing the end ; but it was late in the afternoon before the procession of boats left the ship, under a salute of three rounds of small-arms from

* The Niagara had sixty miles farther to run than the Agamemnon, to land the cable at the head of Trinity Bay.

SECOND EXPEDITION SUCCESSFUL. 187

the detachment of marines on board the Agamemnon, under the command of Lieutenant Morris.

"The progress of the end to the shore was very slow, in consequence of the very stiff wind which blew at the time, but at about three o'clock the end was safely brought on shore at Knigthstown, Valentia, by Mr. Bright and Mr. Canning, the chief and second engineers, to whose exertions the success of the undertaking is attributable, and the Knight of Kerry.* The end was immediately laid in the trench which had been dug to receive it, while a royal salute, making the neighboring rocks and mountains reverberate, announced that the communication between the Old and the New World had been completed."

* A name that occurs several times in this history, and one never to be mentioned but with honor. The Knight of Kerry was Lord of the Isles on that part of the Irish coast; and from the beginning showed the deepest interest in this enterprise; and by his generous hospitality to all connected with it made many friends by whom he was gratefully remembered on both sides of the Atlantic.

CHAPTER XI.

EXCITEMENT IN AMERICA.

WHOEVER shall write the history of popular enthusiasms, must give a large space to the Atlantic Telegraph. Never did the tidings of any great achievement—whether in peace or war—more truly electrify a nation. No doubt, the impression was the greater because it took the country by surprise. Had the attempt succeeded in June, it would have found a people prepared for it. But the failure of the first expedition, added to that of the previous year, settled the fate of the enterprise in the minds of the public. It was a hopeless undertaking; and its projectors shared the usual lot of those who conceive vast designs, and venture on great enterprises, which are not successful, to be regarded with a mixture of derision and pity.

Such was the temper of the public mind, when at noon of Thursday, the fifth of August, the following despatch was received:

"UNITED STATES FRIGATE NIAGARA,
TRINITY BAY, NEWFOUNDLAND, August 5, 1858.

"*To the Associated Press, New York:*

"The Atlantic Telegraph fleet sailed from Queenstown, Ireland, Saturday, July seventeenth, and met in mid-ocean

EXCITEMENT IN AMERICA.

Wednesday, July twenty-eighth. Made the splice at one P.M., Thursday, the twenty-ninth, and separated—the Agamemnon and Valorous, bound to Valentia, Ireland; the Niagara and Gorgon, for this place, where they arrived yesterday, and this morning the end of the cable will be landed.

"It is one thousand six hundred and ninety-six nautical, or one thousand nine hundred and fifty statute, miles from the Telegraph House at the head of Valentia harbor to the Telegraph House at the Bay of Bulls, Trinity Bay, and for more than two thirds of this distance the water is over two miles in depth. The cable has been paid out from the Agamemnon at about the same speed as from the Niagara. The electric signals sent and received through the whole cable are perfect.

"The machinery for paying out the cable worked in the most satisfactory manner, and was not stopped for a single moment from the time the splice was made until we arrived here.

"Captain Hudson, Messrs. Everett and Woodhouse, the engineers, the electricians, the officers of the ship, and in fact, every man on board the telegraph fleet, has exerted himself to the utmost to make the expedition successful, and by the blessing of Divine Providence it has succeeded.

"After the end of the cable is landed and connected with the land line of telegraph, and the Niagara has discharged some cargo belonging to the Telegraph Company, she will go to St. John's for coal, and then proceed at once to New York.

"CYRUS W. FIELD."

The impression of this simple announcment it is impossible to conceive. It was immediately telegraphed to all parts of the United States, and everywhere produced the greatest excitement. In some places all business was suspended; men rushed into the streets, and flocked to the offices where the news was received. At Andover, Massachusetts, the news arrived while the Alumni of the Theological Seminary were celebrating their semi-centennial anniversary by a dinner. One thousand persons were present, all of whom rose to their feet, and gave vent to their excited feelings by continued and enthusiastic cheers. When quiet was restored, Rev. Dr. Adams, of New York, said his heart was too full for a speech, and suggested, as the more fitting utterance of what all felt, that they should join in thanksgiving to Almighty God, and the venerable Dr. Hawes, of Hartford, led them in fervent prayer, acknowledging the great event as from the hand of God, and as calculated to hasten the triumphs of civilization and Christianity. Then all standing up together, sang, to the tune of Old Hundred, the majestic doxology:

> " Praise God, from whom all blessings flow,
> Praise Him all creatures here below;
> Praise Him above, ye heavenly host,
> Praise Father, Son, and Holy Ghost!"

Thus—said Dr. Hawes—" we have now consecrated this new power, so far as our agency is concerned, to the building up of the truth."

In New York the news was received at first with

EXCITEMENT IN AMERICA.

some incredulity. But as it was confirmed by subsequent despatches, the city broke out into tumultuous rejoicing. Never was there such an outburst of popular feeling. In Boston a hundred guns were fired on the Common, and the bells of the city were rung for an hour to give utterance to the general joy. Similar scenes were witnessed in all parts of the United States. I have now before me the New York papers of August, 1858, and from the memorable fifth, when the landing took place, to the end of the month, they contain hardly any thing else than popular demonstrations in honor of the Atlantic Telegraph. It was indeed a national jubilee.

It was natural that this overflow of public feeling should express itself towards one who was recognized as the author of the great work, which inspired such universal joy. Mr. Field, much to his own surprise, "awoke and found himself famous." In twenty-four hours his name was on millions of tongues. Congratulations poured in from all quarters, from mayors of cities and governors of States; from all parts of the Union and the British Provinces; from the President of the United States and the Governor-General of Canada. Mr. Buchanan telegraphed to Mr. Field, at Trinity Bay:

"MY DEAR SIR: I congratulate you with all my heart on the success of the great enterprise with which your name is so honorably connected. Under the blessing of Divine

Providence I trust it may prove instrumental in promoting perpetual peace and friendship between the kindred nations."

The popular estimate of the achievement and its author went still farther. With the natural exaggeration common to masses of men, when carried away by a sudden enthusiasm, the Atlantic Telegraph was hailed as an immense stride in the onward progress of the race, an event in the history of the world hardly inferior to the discovery of America, or to the invention of the art of printing; and the name of its projector was coupled with those of Franklin and Columbus. He who but yesterday was regarded as a visionary, to-day was exalted as a benefactor of his country and of mankind.

This avalanche of praise was quite overwhelming. It is always embarrassing to be forced into sudden conspicuity, and to find one's self the object of general attention and applause. While feeling this embarrassment, Mr. Field could not but be gratified to witness the public joy at the success of the enterprise, and he was deeply touched and grateful for the appreciation of his own services. But probably all these public demonstrations did not go to his heart so much as private letters received from the other side of the Atlantic, from those who had shared the labors of the enterprise—old companions in arms who had borne with him the heavy burden, and now were fully entitled to a share in the honor which was the reward of their common toil.

As a specimen of the congratulations which came from beyond the sea, we quote a single passage from a letter of Mr. George Saward, the Secretary of the Company in London, written immediately on receiving the news of the success of the enterprise. Under the impression of that event, he writes to Mr. Field:

"At last the great work is successful. I rejoice at it for the sake of humanity at large. I rejoice at it for the sake of our common nationalities, and last but not least, for your personal sake. I most heartily and sincerely rejoice with you, and congratulate you, upon this happy termination to the trouble and anxiety, the continuous and persevering labor, and never-ceasing and sleepless energy, which the successful accomplishment of this vast and noble enterprise has cost you. Never was man more devoted—never did man's energy better deserve success than yours has done. May you in the bosom of your family reap those rewards of repose and affection, which will be doubly sweet from the reflection, that you return to them after having been under Providence the main and leading principal in conferring a vast and enduring benefit on mankind. If the contemplation of fame has a charm for you, you may well indulge in the reflection; for the name of Cyrus W. Field will now go onward to immortality, as long as that of the Atlantic Telegraph shall be known to mankind."

The Directors, whose faith and courage had been so severely tried, now felt double joy, for their friend and for themselves, at this glorious result of their united labors. Mr. Peabody wrote to Mr. Field that "his reflections must be like those of Columbus, after the discovery of America." Sir Charles Wood and Sir John Pakington, who, as successive First Lords of the Admiralty, had supported the enterprise with the constant aid of the British Government, wrote letters of congratulation on the great work which had been carried through mainly by his energy and unconquerable will. They were above any petty national jealousy, and never imagined that it would detract aught from the just honor of England, to award full praise to the courage and enterprise of an American.

On his part, Mr. Field was equally anxious to acknowledge the invaluable aid given by others—aid, without which the efforts of no single individual could command success. On his arrival at St. John's, he was welcomed with enthusiasm by the whole population. An address was presented to him by the Executive Council of Newfoundland, in which they offered their hearty congratulations on the success of the undertaking, which they recognized as chiefly due to him. "Intimately acquainted as we have been"—these are their words—"with the energy and enterprise which have distinguished you from the commencement of the great work of telegraph connection between the Old and the New Worlds; and feeling that under Providence this triumph of science is mainly due to your well-directed and indomitable exertions, we desire to express to you our high appreciation of your success in the cause of the world's progress," etc.; to which Mr. Field replied, recognizing in turn the cordial support which he had always received from the Government of Newfoundland. The Chamber of Commerce of St. John's also presented an address in similar terms, to which he replied—after acknowledging their kind mention of his own labors and sacrifices:

"But it would not only be ungenerous, but unjust, that I should for a moment forget the services of those who were my co-workers in this enterprise, and without whom any labors of mine would have been unavailing. It would be difficult to enumerate the many gentlemen whose scientific acquirements and skill and energy have been devoted to the advancement of this work, and who have so mainly produced the issue which has called forth this expression of your good wishes on my behalf. But I could not do justice to my own feelings did I fail to acknowledge how much is owing to Captain Hudson and the officers of the Niagara, whose hearts were in the work, and whose toil was unceasing; to Captain Dayman of her Majesty's ship Gorgon, for the soundings so accurately made by him last year, and for the perfect manner in which he led the Niagara over the great-circle arc while laying the cable; to Captain Otter, of the Porcupine, for the careful survey made by him in Trinity Bay, and for the admirable manner in which he piloted the Niagara at night to her anchorage; to Mr. Everett, who has for months devoted his whole time to designing and perfecting the beautiful machinery that has so successfully paid out the cable from the ships—machinery so perfect in every respect, that it was not for one moment stopped on board the Niagara until she reached her destination in Trinity Bay; to Mr. Woodhouse, who superintended the coiling of the cable, and zealously and ably coöperated with his brother engineer during the progress of paying-out; to the electricians for their constant watchfulness ; to the men for their almost ceaseless labor (and I feel confident that you will have a good report from the commanders, engineers, electricians, on board the Agamemnon and Valorous, the Irish portion of the fleet); to the Directors of the Atlantic Telegraph Company for the time they have devoted to the undertaking without receiving any compensation for their services (and it must be a pleasure to many of you to know that the director, who has devoted more time than any other, was for many years a resident of this place, and well known to all of you—I allude to Mr. Brooking, of London); to Mr. C. M. Lampson, a native of New England, but who has for the last twenty-seven years resided in London, who appreciated the great importance of this enterprise to both countries, and gave it most valuable aid, bringing his sound judgment and great business talent to the service of the Company; to that distinguished American, Mr. George Peabody, and his worthy partner, Mr. Morgan, who not only assisted it most liberally with their means, but to whom I could always go with confidence for advice."

Such acknowledgments, constantly repeated, showed a mind incapable of envy or jealousy ; that was chiefly anxious to recognize the services of others, and that they should receive from the public, both of England

and America, the honors which they had so nobly earned.

After two or three days' delay at St. John's, which the Niagara was obliged to make for coal, but which the people spent in festivity and rejoicing, she left for New York, where she arrived on the eighteenth—two weeks from the landing of the cable in Trinity Bay. These had been weeks of great excitement, yet not unmingled with suspense and anxiety. The public, eager for news, devoured every thing that concerned the telegraph with impatience, but all was not satisfactory. Despatches from Trinity Bay said that signals were continually passing over the cable, yet no news reached the public from the other side of the Atlantic. This was partially explained by a message from Mr. Field, sent from Trinity Bay to the Associated Press as early as the seventh:

"We landed here in the woods, and until the telegraph instruments are perfectly adjusted, no communications can pass between the two continents; but the electric currents are received freely.

"You shall have the earliest intimation when all is ready, but it may be some days before every thing is perfected. The first through message between Europe and America will be from the Queen of Great Britain to the President of the United States, and the second his reply"

But as the public grew impatient, and friends sent anxious inquiring messages, he telegraphed again from St. John's on the eleventh:

"Before I left London, the Directors of the Atlantic Telegraph Company decided unanimously that, after the cable was laid, and the Queen's and President's messages transmitted, the line should be kept for several weeks for the sole use of Dr. Whitehouse, Professor Thomson, and other electricians, to enable them to test thoroughly their several modes of telegraphing, so that the Directors might decide which was the best and most rapid method for future use; for it was considered that after the line should be once thrown open for business, it would be very difficult to obtain it for experimental purposes, even for a short time.

"Due notice will be given when the line will be ready for business, and the tariff of prices."

Still the public were not satisfied, and many were beginning to doubt, when, on the sixteenth, it was suddenly announced that the Queen's message was received. It was as follows:—

"*To the President of the United States, Washington:*

"The Queen desires to congratulate the President upon the successful completion of this great international work, in which the Queen has taken the deepest interest.

"The Queen is convinced that the President will join with her in fervently hoping that the electric cable which now connects Great Britain with the United States will prove an additional link between the nations, whose friendship is founded upon their common interest and reciprocal esteem.

"The Queen has much pleasure in thus communicating with the President, and renewing to him her wishes for the prosperity of the United States."

To this the President replied:

"WASHINGTON CITY, August 16, 1858.
"*To Her Majesty Victoria, the Queen of Great Britain:*
"The President cordially reciprocates the congratulations of her Majesty the Queen, on the success of the great international enterprise accomplished by the science, skill, and indomitable energy of the two countries.

"It is a triumph more glorious, because far more useful to mankind, than was ever won by conqueror on the field of battle.

"May the Atlantic Telegraph, under the blessing of Heaven, prove to be a bond of perpetual peace and friendship between the kindred nations, and an instrument destined by Divine Providence to diffuse religion, civilization, liberty, and law throughout the world.

"In this view, will not all nations of Christendom spontaneously unite in the declaration that it shall be for ever neutral, and that its communications shall be held sacred in passing to their places of destination, even in the midst of hostilities?
"JAMES BUCHANAN."

The arrival of the Queen's message was the signal for a fresh outbreak of popular enthusiasm. The next morning, August seventeenth, the city of New York was awakened by the thunder of artillery. A hundred guns were fired in the City Hall Park at daybreak, and the salute was repeated at noon. At this hour, flags were flying from all the public buildings, and the bells of the principal churches began to ring, as Christmas bells signal the birth of one who came to bring peace and good-will to men—chimes that, it was fondly hoped, might usher in a new era, as they should

> Ring out the old, ring in the new,
> Ring out the false, ring in the true.

That night the city was illuminated. Never had it seen so brilliant a spectacle. Such was the blaze of light around the City Hall, that the cupola caught fire, and was consumed, and the Hall itself narrowly escaped destruction. Similar demonstrations took place in other parts of the United States. From the Atlantic to the Valley of the Mississippi, and to the Gulf of Mexico, in every city was heard the firing of guns and the ringing of bells. Nothing seemed too extravagant to give expression to the popular rejoicing.

The next morning after this illumination, the Niagara entered the harbor of New York, and Mr. Field hastened to his home. The night before leaving the ship, he had written to the Directors in London, giving a full report of the laying of the cable, which he closed by resigning the position which he had held for the last seven months. He wrote:

"At your unanimous request, but at a very great personal sacrifice to myself, I accepted the office of General Manager of the Atlantic Telegraph Company, for the sole purpose of doing all in my power to aid you to make the enterprise successful; and as that object has been attained, you will please accept my resignation. It will always afford me pleasure to do any thing in my power, consistent with my duties to my family and my own private affairs, to promote the interests of the Atlantic Telegraph Company."

Once more with his family, Mr. Field hoped for a brief interval of rest and quiet. But this was impossible. The great event with which his name was connected was too fresh in the public mind. He could not escape public observation. He was at once thronged with visitors, offering their congratulations, and his house surrounded with crowds eager to see and hear him. While making all allowance for popular excitement, yet none could deny that a service so great demanded some public recognition. Even in England, where the enthusiasm did not approach that in this country, still the wondrous character of the achievement was fully acknowledged. Said the London Times on the morning of the sixth of August: "Since the discovery of Columbus, nothing has been done in any degree comparable to the vast enlargement which has thus been given to the sphere of human activity." "More was done yesterday for the consolidation of our empire, than the wisdom of our statesmen, the liberality of our Legislature, or the loyalty of our colonists, could ever have effected." To mark the public benefit which had been conferred, the Chief Engineer of the Expedition, Mr. Charles T. Bright, was knighted, and Captains Preedy and Aldham were both made Companions of the Bath, and other officers were promoted. Thus England showed her appreciation of their services.

But in this country titles and honors come not from

the Government, but from the people. Popular enthusiasm exhausted itself in eulogies of the man who had linked the Old World to the New. It seems strange now to sit down in cold blood and read what was published in the papers of that day. A collection of American journals issued during that eventful month, August, 1858, would be a literary curiosity.*

* Such a curiosity exists, prepared by the industry of a gentleman who was one of the most careful collectors of the events of his time— by which he gathered up the materials of future history—Mr. John R. Bartlett, formerly Secretary of State of Rhode Island. This gentleman kept files of all the papers referring to the Atlantic Telegraph, from which he compiled a very unique volume. It is in the form of a scrap-book, but on a gigantic scale, being of a size equal to Webster's large Dictionary. It is made up entirely of newspaper cuttings, classified under different heads, and neatly arranged in double columns on nearly four hundred folio pages. The matter thus compressed would make between three and four octavo volumes of the size of Prescott's Histories, if printed in the style of those works. Every thing is included that could be gathered from European as well as American papers, touching the claims of the inventors and projectors of the electric telegraph in general, and of the Atlantic Telegraph in particular. The historical sketches are set off by illustrations taken from the pictorial papers. Altogether it embraces more of the materials of a history of this subject than any other volume with which we are acquainted, and well deserves the title prefixed to it by the laborious compiler:

"THE ATLANTIC TELEGRAPH.—Its Origin and History, with an Account of the Voyages of the Steamers Niagara and Agamemnon, in Laying the Cable, and of the Celebration of the Great Event in New York, Philadelphia, Brooklyn, Montreal, Dublin, Paris, etc.; together with the Discussions, Sermons, Poetry, and Anecdotes relating thereto; also, a History of the Invention of the Electric Telegraph. Illustrated with Maps, Plans, Views, and Portraits, collected from the Newspapers of the Day, and arranged by John Russell Bartlett. 1858."

Nor was it merely in such outward demonstrations that the public enthusiasm showed itself. The feeling struck deeper, and reached all minds. While the people shouted and cannon roared, sober and thoughtful men pondered on the change that was being wrought in the earth. Business men reasoned how it would affect the commerce of the world, while the philanthropic regarded it as the forerunner of an age of universal peace. The first message flashed across the sea—even before that of the Queen—had been one of religious exultation. It was from the Directors in Great Britain to those on this side the Atlantic, and, simply reciting the fact that Europe and America were united by telegraph, at once broke into a strain of religious rapture, echoing the song of the angels over a Saviour's birth: "Glory to God in the highest; on earth, peace, good-will toward men." Poetry at once caught up the strain. The event became the theme of innumerable odes and hymns, of which it must be said that, whatever their merit as poetry, their spirit at least was noble, celebrating the event chiefly as promoting the brotherhood of the human family. The key-note was struck in such lines as these:

> 'Tis done! the angry sea consents,
>   The nations stand no more apart,
> With claspèd hands the continents
>   Feel throbbings of each other's heart.

> Speed, speed the cable; let it run
> A loving girdle round the earth,
> Till all the nations 'neath the sun
> Shall be as brothers of one hearth;
>
> As brothers pledging, hand in hand,
> One freedom for the world abroad,
> One commerce over every land,
> One language and one God.

The sermons preached on this occasion were literally without number. Enough found their way into print to make a large volume. Never had an event touched more deeply the spirit of religious enthusiasm. Devout men held it as an advance toward that millennial era which was at once the object of their faith and hope. Was not this the predicted time when, "many should run to and fro, and knowledge should be increased?" So said the preachers, taking for their favorite text the vision of the Psalmist, "Their line is gone out through all the earth, and their words to the end of the world;" or the question of Job: "Canst thou send forth the lightnings, that they may go and say unto thee, Here we are?" Was not this the dawn of that happy age, when all men should be bound together in peaceful intercourse, and nations should learn war no more? Such was the burden of the discourses that were preached in a thousand pulpits from one end of the country to the other. Even the Roman Catholic Church, so lofty and inflexible in its claims, soaring into the past centuries, and almost disdaining the material progress of the present day as compared with the spiritual glories of the Ages of Faith, did not ignore the great event; and in laying the foundation of the new Cathedral of St. Patrick, the largest temple of religion on the continent, Archbishop Hughes placed under the corner-stone an inscription, wherein, along with the enduring record of the Christian faith and the names of martyrs and confessors, he did not disdain to include a brief memorial of this last achievement of science, and the name of him who had conferred so great a benefit on mankind.

These public demonstrations culminated on the first of September, when the city authorities gave a public ovation to Mr. Field and the officers of the expedition. In accepting these honors, Mr. Field had taken good care that the British officers should be included with the American. At St. John's he had been notified of the intended celebration, and at once telegraphed to the British Admiral at Halifax:

"I should consider it a very great personal favor if you would permit the Gorgon, Captain Dayman, to accompany the Niagara, Captain Hudson, to New York. English officers and English sailors have labored with American officers and American sailors to lay the Atlantic cable. They were with us in our days of trial, and pray let them, if you can, share with us our triumph."

The request was granted so far as this, that the officers were allowed leave of absence, and came on to New York to take part in the celebration, and in all the honors which followed, the officers of the Gorgon were associated with those of the Niagara.

The day arrived, and the celebration surpassed any thing which the city had ever witnessed before. It was a mild autumn day—warm, yet with a sky softly veiled with clouds, that seemed to invite a whole population into the streets. The day commenced with a solemn service at Trinity Church, which was attended by the city authorities, the representatives of foreign powers, and an immense concourse of people. The vast edifice was decorated with evergreens; in the centre hung a cross, with the inscription: "Glory to God on high; and on earth, peace, good-will towards men." When the audience were assembled, there entered a procession of two hundred clergy, headed by Bishop Doane of New Jersey, who was to deliver the address. Prayers were offered and Scriptures were read, and at intervals the choir gave voice to the general joy in the anthems in which for ages the Church has been wont to pour forth its exultation: "O come, let us sing unto the Lord," the Gloria in Excelsis, and the Te Deum Laudamus.

At noon, Mr. Field and the officers of the ships landed at Castle Garden and were received with a national salute. A procession was formed which extended for miles from the Battery to the Crystal Palace, which stood on the plot of ground now known as Bryant Park, between Fortieth and Forty-second streets. In the procession were Lord Napier, the British Minister, and officers of the army and navy. For the whole distance the streets were crowded. The windows and even the tops of the houses were filled with people. Everywhere flags and banners, with every device, floated in the air. So dense was the crowd that it was five or six hours before the procession could reach the Crystal Palace.

Here its coming was awaited by an assembly that filled all the aisles and galleries. An address was delivered, giving the history of the Atlantic Telegraph. The Mayor then rose, and presenting Mr. Field to the audience, spoke as follows:

"SIR : History records but few enterprises of such 'pith and moment' as to command the attention and at the same time enlist the sympathies of all mankind. In all ages warlike expeditions have been undertaken on a scale of grandeur sufficient to astonish the world ; but the evils which are inseparable from their prosecution have always sent a thrill of horror through the anxious nations. The discovery of the Western continent even, the grandest event of modern times, was made by an insignificant fleet which left the shores of Spain without attracting the notice of the civilized world. Far different has been the history of the daring and difficult enterprise of uniting the Old World and the New by means of the electric telegraph.

From the very outset the good, the great and the wise of all lands beneath the sun, have watched with intense anxiety, and even when doubt existed, with warm interest, every step taken toward the accomplishment of what was universally acknowledged to be the most momentous undertaking of an age made marvellous by wonderful scientific and mechanical achievements. The two greatest and freest nations of the globe, by independent constitutional legislation, and by the aid of their finest ships and their ablest officers and engineers, combined together to insure success. Capital was liberally subscribed by private citizens in a spirit which put greed to the blush. The press on both sides of the Atlantic recorded the details of the progress of the undertaking with cordial interest, and secured the generous sympathies of men of all kindreds and tongues and nations in its behalf. You were thus fortunate, sir, in being identified with a project of such magnificent proportions and universal concern. But the enterprise itself was no less fortunate in being projected and carried into execution by a man whom no obstacles could daunt, no disasters discourage, no doubts paralyze, no opposition dishearten. If you, to whom the conduct of this great enterprise was assigned by the will of Providence and the judgment of your fellow-men, had been found wanting in courage, in energy, in determination, and in a faith that was truly sublime, the very grandeur of the undertaking would only have rendered its failure the more conspicuous. But, sir, the incidents of the expedition, and the final result—too familiar to all the world to need repetition here—have demonstrated that you possessed all the qualities essential to achieve a successful issue. It is for this reason that you now stand out from among your fellow-men a mark for their cordial admiration and grateful applause. The city of your home delights to honor you; your fellow-citizens, conscious that the glory of your success is reflected back upon them, are proud that your lot has been cast among them. They have already testified their appreciation of your great services and heroic perseverance by illuminations, processions, serenades, and addresses. And now, sir, the municipal government of this, the first city on the Western continent, instruct me, who have never felt the honor of being its chief magistrate so sensibly as in the presence of this vast assemblage of its fair women and substantial citizens, to present to you a gold box, with the arms of the city engraved thereon, in testimony of the fact that to you mainly, under Divine Providence, the world is indebted for the successful execution of the grandest enterprise of our day and generation; and in behalf of the Mayor, Aldermen, and Commonalty of the City of New York, I now request your acceptance of this token of their approbation. In conclusion, sir, of this, the most agreeable duty of my public life, I sincerely trust that your days may be long in the land, and as prosperous and honorable as your achievement in uniting the two hemispheres by a cord of electric communication has been successful and glorious."

To this flattering address, Mr. Field replied :

"Sir : This will be a memorable day in my life; not only because it celebrates the success of an achievement with which my name is connected, but because the honor comes from the city of my home—the metropolitan city of the new world. I see here not only the civic authorities and citizens at large, but my own personal friends—men with whom I have been connected in business and friendly intercourse for the greater part of my life. Five weeks ago, this day and hour, I was standing on the deck of the Niagara in mid-ocean, with the Gorgon and Valorous in sight, waiting for the Agamemnon. The day was cold and cheerless, the air was misty, and the wind roughened the sea; and when I thought of all that we had passed through—of the hopes thus far disappointed, of the friends saddened by our reverses, of the few that remained to sustain us—I felt a load at my heart almost too heavy to bear, though my confidence was firm, and my determination fixed. How different is the scene now before me—this vast crowd testifying their sympathy and approval, praises without stint, and friends without number! This occasion, sir, gives me the opportunity to express my thanks for the enthusiastic reception which I have received, and I here make my acknowledgments before this vast concourse of my fellow-citizens. To the ladies I may, perhaps, add, that they have had their appropriate place, for when the cable was laid, the first public message that passed over it came from one of their own sex. This box, sir, which I have the honor to receive from your hand, shall testify to me and to my children what my own city thinks of my acts. For your kindness, sir, expressed in such flattering, too flattering terms, and for the kindness of my fellow-citizens, I repeat my most heartfelt thanks."

The enthusiasm with which this address was received reached its height, when at the close, Mr. Field advanced to the edge of the platform, and unrolling a despatch, held it up, saying: "Gentlemen, I have just received a telegraphic message from a little village, now a suburb of New York, which I will read to you:

"London, September 1, 1858.

"To Cyrus W. Field, New York:

"The directors are on their way to Valentia, to make arrangements for opening the line to the public. They convey, through the cable, to you and your fellow-citizens, their hearty congratulations and good wishes, and cordially sympathize in your joyous celebration of the great international work."*

A gold medal was presented to Captain Hudson, with an address, to which he made a fitting reply. Similar testimonials were presented to all the English captains through Mr. Archibald, the British Consul, who replied for his absent countrymen, after which the whole audience rose to their feet, as the band played "God save the Queen."

It was long after dark when the exercises closed, and the vast multitude dispersed.

The night witnessed one of those displays for which New York surpasses all the cities of the world—a firemen's torchlight procession—a display such as was afterward given to the Prince of Wales, but which we

---

* The history of this despatch is curious. Though dated at London, it was sent from a small town in Ireland. The directors were on their way from Dublin to Valentia, on the morning of the first of September, when Mr. Saward remarked: "This is the day of the celebration in New York—we ought to send a despatch to Mr. Field." Accordingly, at the first stopping-place (Mallow Station) the message was written, and forwarded to Valentia, and thence sent across the Atlantic. It was put into Mr. Field's hand as he was getting into his carriage on the Battery.

shall probably witness no more, since the Volunteer Fire Department is disbanded.

But one day did not exhaust the public enthusiasm. The next evening, a grand banquet was given by the city authorities, at which were present a great number of distinguished guests. Lord Napier spoke, in language as happy as it was eloquent, of the new tie that was formed between kindred dwelling on opposite sides of the sea, and awarded the highest praise to the one whom he recognized as the author of this great achievement.

While these demonstrations continued, every opposing voice was hushed in the chorus of national rejoicing; yet some there were, no doubt, who looked on with silent envy or whispered detraction. But who could grudge these honors to the hero of the hour—honors so hardly won, and which, as it proved, were soon to give place to harsh censures and unjust imputations?

Alas for all human glory! Its paths lead but to the grave. Death is the end of human ambition. The very day that a whole city rose up to do honor to the Atlantic Telegraph and its author, it gave its last throb, and that first cable was thenceforth to sleep for ever silent in its ocean grave.

CHAPTER XII.

DID THE FIRST CABLE EVER WORK?

THE Atlantic cable was dead! That word fell heavy as a stone on the hearts of those who had staked so much upon it. What a bitter disappointment to their hopes! In all the experience of life there are no sadder moments than those in which, after years of anxious toil, striving for a great object, and after one glorious hour of triumph, the achievement that seemed complete becomes a total wreck. Vain is all human toil and endeavor. The years thus spent are fled away; the labor that was to have brought such a reward of "riches and honor," is lost; and the prolonged tension of the mind by the excitement of hope and ambition, and the temporary success, reacts to plunge it into a deeper depression. So was it here. Years of labor and millions of capital were swept away in an hour into the bosom of the pitiless sea.

Of course the reaction of the public mind was very great. As its elation had been so extravagant before, it was now silent and almost sullen. People were ashamed of their late enthusiasm, and disposed to revenge themselves on those who had been the objects of their idolatry. It is instructive to read the papers of the day. As soon as it was evident that the Atlantic cable was a dead lion, many hastened to give it a parting kick. There was no longer any dispute as to who was the author of the great achievement. Rival claimants quietly withdrew from the field, content to leave him alone in his glory.

Many explanations were offered of this sudden suspension of life. One writer argued that the Telegraphic Plateau was only a myth; that the bottom of the ocean was jagged and precipitous; that the cable passed over lofty mountain chains, and hung suspended from the peaks of submarine Alps, till it broke and fell into the tremendous depths below.

But others found a readier explanation. With the natural tendency of a popular excitement to rush from one extreme to the other, many now believed that the whole thing was an imposition on public credulity, a sort of "Moon hoax." An elaborate article appeared in a Boston paper, headed with the alarming question, "Was the Atlantic cable a humbug?" wherein the writer argued through several columns that it was a huge deception. A writer in an English paper also made merry of the celebration in Dublin, where a banquet was given to Sir Charles Bright, in an article bearing the ominous title: "Very like a whale!" This writer proved not only that the Atlantic cable was never laid, but that such a thing was mathematically

DID THE FIRST CABLE EVER WORK? 215

impossible. But since he turned out to be a crazy fellow, whom the police had to take into custody, his "demonstrations" did not make much impression on the public. The difficulty of finding a motive for the perpetration of such a stupendous fraud, did not at all embarrass these ingenious writers. Was it not enough to make the world stare? to furnish something to the gaping crowd, even though it were but a nine days' wonder? Those who thus reasoned seemed not to reflect that such deceptions are always sure to be found out; that one who goes up like a rocket comes down like a stick; and that if by false means he has made himself an object of popular idolatry, he is likely to become the object of popular indignation.

But others there were—sharp, shrewd men—who thought they could see through a mill-stone farther than their neighbors, who shook their heads with a knowing air, and said: "It was all a stock speculation." One writer stepped before the public with this solemn inquiry: "Now that the great cable glorification is over, we should like to ask one question: How many shares of his stock did Mr. Field sell during the month of August?" This he evidently thought was a question which could not be answered, except by acknowledging a great imposition on the public. If this brilliant inquirer after truth really desired to be informed, we could have referred him to Messrs. George Peabody & Co., of London, with whom was deposited all of Mr. Field's stock at the time, and who, during that memorable month of August, sold *just one share*, and that at a price below the par value, which had been paid by Mr. Field himself. Whether this was an object sufficiently great to set two hemispheres in a blaze, we leave him to judge.

To those who have followed this narrative, all these conjectures and suspicions will appear very absurd. The personal reflections of course deserved and received only the contempt with which a man of character always scorns an imputation on his personal honor. But while these anonymous scribblers might be despised, many honest people not disposed to think evil were sorely perplexed. That the cable should continue to work for three or four weeks, *and then stop the very day of the celebration*, was certainly a singular, if not a suspicious circumstance; and it was not to be wondered at that it should excite a painful feeling of doubt. The distrust is quite natural, and ought not to be matter either of offence or surprise. On the contrary, those who are fully satisfied of the facts, ought rather to be glad of the opportunity which such questions afford, to present the amplest vindication.

To relieve all doubts, it is only necessary to give a very brief history of the working of the Atlantic cable. It was landed on both sides of the ocean on the fifth of August. The last recorded message passed over it

DID THE FIRST CABLE EVER WORK? 217

on the first of September, one day short of four weeks. Within that time there were sent exactly four hundred messages, of which two hundred and seventy-one were from Newfoundland to Ireland, and one hundred and twenty-nine from Ireland to Newfoundland. Of these, the greater part were merely between the operators themselves, respecting the adjustment of instruments, and working the telegraph, which, while they furnished decisive evidence *to them*, were of no force to the public. Of course an operator, working with a battery on the shore at Valentia, or at Trinity Bay, watching his instrument, and seeing the little tongue of light reflected from the moving mirror of the galvanometer, needed no other evidence of an electric current that had passed through the cable. He *saw* it, and knew, as if he saw the flash of a gun on the coast of Ireland, that it was a light which had come from beyond the sea. But these private assurances were nothing to the outside world. What they needed was *public* messages, conveying news from one hemisphere to the other. Of these, there were not a great number, for obvious reasons. The cable, during the four weeks of its existence, never worked *perfectly*—that is, as a land line works, transmitting messages freely and rapidly, and with perfect accuracy. It was subject to frequent interruptions for reasons which may satisfy any one that the wonder is, not that it did so little, but that it did so much.

1st. To begin with, the cable was not constructed in

the most perfect manner. Its makers, though the best then in the world, had had but little experience in making deep-sea cables. No line over three hundred miles long had ever been laid. 2d. It had been made more than a year before. After it was finished, part of it had been coiled out of doors, where it was exposed to a burning sun, by which, as was afterward found, the gutta-percha had been melted in many places till the insulation was nearly destroyed. 3d. It had been put on board the ships in 1857, and after the first failure, had been taken out and coiled on the dock at Plymouth, and then re-shipped in 1858. Thus it had been twisted and untwisted, some portions of it as many as ten times. Then the Agamemnon was so shaken in the terrible gale of June, that the cable on board of her was seriously injured, and some portions were cut out and condemned. Taking all these things together, the wonder is, not that the cable failed after a month, but that it ever worked at all!

Owing to this impaired state of the cable, it did *not* work perfectly. Probably it would not have worked at all with ordinary instruments. But the galvanometer of Professor Thomson, that instrument of marvellous delicacy, drew faint whispers from its muttering lips. Signals came and went, which showed that the electric current passed from shore to shore, and gave promise that with delicate handling it could be taught to speak plainly. But for the present it spoke slowly and with difficulty. It often took hours to get through a single despatch of any length. Witness the delay in transmitting the Queen's message! These frequent interruptions were ascribed to various causes. Sometimes it was earth-currents; at others, a thunderstorm was raging. Thus, on the morning of Thursday, the twenty-sixth of August, there was a violent storm in Newfoundland, heavy rain, accompanied by thunder and lightning. At three o'clock, the lightning was so intense that for an hour and a half the end of the cable had to be put to the earth for protection. After that the storm cleared away, and at seven o'clock the weather was reported as very fine. But aside from these local and temporary causes, the real difficulty was in the cable itself, whose insulation had been fatally impaired, and which was now wearing out its life on the rocks of the sea. These causes made its speech difficult and broken. Yet sometimes it flashed up with sudden power. In one case, a message was sent from the office at Trinity Bay to Ireland and an answer received back in two minutes! Such incidents excited the liveliest hopes that all difficulties would be speedily overcome, and justified the messages which were sent to the New York papers from day to day, that the instruments were being adjusted, by which it was expected that the line would soon be put in perfect working order, and be thrown open to the public.

But these flashes of light proved to be only the flickering of the flame, that was soon to be extinguished in the eternal darkness of the waters.

But the question which perplexed not only skeptics, but the truest friends, was not whether the cable worked fast or slow, *but whether it ever worked at all.* Happily, this is a question which can easily be settled, since it is one simply of facts and dates, which can be ascertained by referring to the files of the English and American papers. Of course the only proof must be in messages containing *news*. Mere congratulations between the Queen and the President, or the Mayor of New York and the Mayor of London, prove nothing, for these might have been prepared beforehand, if we suppose a design to impose on the credulity of the public. But the decisive test is this: Was there at any time within that month published in the English or American journals *news* which could not be matter of guess or conjecture, and within a time too short for its possible transmission in any other way? If this can be proved beyond all doubt, even in a few instances, the question is decided, for the argument is just as strong with a dozen cases as with a thousand. We give, therefore, a few dates, the accuracy of which can be tested by any one who will take the trouble to examine the English and American papers:

On Saturday, the fourteenth of August, the steam-

DID THE FIRST CABLE EVER WORK? 221

ships Arabia and Europa, the former bound for New York and the latter for Liverpool, came into collision off Cape Race. The accident was not known in New York until Tuesday, the seventeenth, since it could not be telegraphed till the Arabia reached Halifax or the Europa St. John's, into which port she put for repairs. As soon as the news reached New York, the agent of the Company, Mr. Nimmo (Mr. Cunard himself being then in England), at once prepared a despatch to be sent to relieve immediate anxiety. This was not forwarded to Newfoundland, as peremptory orders had been given not to transmit any private business messages to go through the cable until the line was fully open to the public. But the next day Mr. Field arrived in New York, and Mr. Nimmo applied to him. Seeing the urgency of the case, he ordered it to be forwarded. It was accordingly sent, and arrived in London on the twentieth, giving the first news that was received of the accident. This was repeatedly stated by the late Sir Samuel Cunard, of London, and confirmed by his son Mr. Edward Cunard, of New York. The message was published in the London papers of the twenty-first, as follows:

"Arabia in collision with Europa, Cape Race, Saturday. Arabia on her way. Head slightly injured. Europa lost bowsprit, cutwater stem sprung. Will remain in St. John's ten days from sixteenth. Persia calls at St. John's for mails and passengers. No loss of life or limb."

This first news message was not only a very decisive one as to the fact of telegraphic communication, but one which showed the relief given by speedy knowledge in dispelling doubt and fear. Mr. William E. Dodge, of New York, says: "I was in Liverpool at the time, and expecting friends by the Europa. Any delay in the arrival of the ship would have caused great anxiety. But one morning, on going down to the Exchange, we saw posted up this despatch received the night before by the Atlantic Telegraph. All then said, if the cable never did any thing more, it had fully repaid its cost." Well may he add with devout feeling: "It seemed as if Divine Providence had permitted the event, to furnish a testimony which could not be denied, to the reality and the benefit of this new means of communication between the two continents."

Passing over all the messages exchanged between the operators at the stations, the congratulations of Queen and President, and of the Mayors of New York and London, we come to another news despatch. August twenty-fifth, Newfoundland reports to Valentia:

"Persia takes Europa's passengers and mails. Great rejoicing everywhere at success of cable. Bonfires, fireworks, *feux de joie*, speeches, balls, etc. *Mr. Eddy, the first and best telegrapher in the States, died to-day.* Pray give some news for New York; they are mad for news."

This despatch the writer, who was then in Europe, read first in the London Times. The item which ar-

DID THE FIRST CABLE EVER WORK ?

rested his attention was the death of Mr. Eddy, as he had some acquaintance with that gentleman.

That the news must have come by cable, is clearly shown by an examination of dates. He died suddenly, at Burlington, Vermont, Monday, August twenty-third, 1858, at ten o'clock fifteen minutes A. M. The exact day and hour we learned from his widow, who after his death lived in Brooklyn. The news was telegraphed to New York, and from there sent to Trinity Bay, which it reached the following day, and from which it was forwarded to Valentia, and appeared in the London Times Wednesday morning. Thus not forty-eight hours elapsed after he breathed his last, before it was published in England. If any one wishes to see the despatch, he will find a file of The Times in the Astor Library.

But here appears a slight discrepancy, that, however, when examined, furnishes double proof. The despatch is dated August twenty-fifth, and says Mr. Eddy died *to-day*, and yet it is published in the London Times of the same date! How is this? It was sent between nine and ten o'clock at night of the twenty-fourth, when the operator at Heart's Content would say *this day* of a piece of news just received, but in affixing the date, he was governed *by Greenwich time*, which made it more than three hours later. Accordingly it was published in The Times, dated August twenty-fifth, fifty-three minutes past twelve A. M. !

Those who argued for the theory of collusion and deception, must have been embarrassed by this unexpected intelligence appearing in London, which could only be explained as a false report, unless (more wonderful still!) Mr. Eddy had entered into the plot, and sent the message beforehand, and then offered himself as a sacrifice, to prove it correct!

To the demand for news in the above despatch, a reply was at once returned: "Sent to London for news." And later the same day came the following:

"North American with Canadian, and the Asia with direct Boston mails, leave Liverpool, and Fulton, Southampton, Saturday next. To-day's morning papers have long, interesting reports by Bright. Indian news. Virago arrived at Liverpool to-day; Bombay dates nineteenth July. Mutiny being rapidly quelled."

A despatch of the same date, August twenty-fifth, also announces peace with China. The whole was received at Trinity Bay about nine o'clock P.M., and would have been sent on at once to New York, but that the land lines in Nova Scotia were closed at that hour. It was sent the next morning, and appeared in the evening papers of the twenty-sixth.

By referring again to the London Times, the reader will see that the news from China was published in London on the twenty-third of August. It was there given as *unexpected news*, so that it could not have been a shrewd guess on the part of anybody

DID THE FIRST CABLE EVER WORK? 225

either in England or America. It took the public by surprise, both for the news itself and *for the way in which it came*—which was not by India and the Red Sea, but by St. Petersburg, where it arrived on the twenty-first, having been brought overland by a courier to Prince Gortchakoff. From there it was telegraphed to the Government at Paris, and thence to London. The Times comments on this roundabout way in which intelligence so important reached England. Yet this news, so unlooked for, announced in London only on the morning of the twenty-third of August, was published in New York on the twenty-sixth.

August twenty-seventh, comes a still longer despatch, which we give in full:

"George Saward, Secretary Atlantic Telegraph Company, to Associated Press, New York. News for America by Atlantic cable. Emperor of France returned to Paris, Saturday. King of Prussia too ill to visit Queen Victoria. Her Majesty returns to England thirtieth of August.—St. Petersburg, twenty-first of August. Settlement of Chinese question. Chinese empire opened to trade; Christian religion allowed; foreign diplomatic agents admitted; indemnity to England and France.—Alexandria, August ninth. The Madras arrived at Suez seventh inst. Dates Bombay to the nineteenth; Aden, thirty-first. Gwalior insurgent army broken up. All India becoming tranquil."

This despatch embodies about a dozen distinct items of news, not one of which could be known without a telegraphic communication. The whole was received in New York, and published in the evening papers *the same day.*

Not to be outdone in giving news, the next day, Saturday, August twenty-eighth, Newfoundland thus replies to Valentia:

"To the Directors: Take news first, Saward. Sir William Williams, of Kars, arrived Halifax Tuesday. Enthusiastically received. Immense procession—welcome address —feeling reply. Held levee—large number presented. Niagara sailed for Liverpool at one this morning. The Gorgon arrived at Halifax last night. Yellow fever in New Orleans, sixty to seventy deaths per day. Also declared epidemic, Charleston. Great preparations in New York and other places for celebration, to be held the first and second of September. New Yorkers will make it the greatest gala-day ever known in this country. Hermann sailed for Fraser's River; six hundred passengers. Prince Albert sailed yesterday for Galway. Arabia and Ariel arrived New York; Anglo Saxon, Quebec; Canada, Boston. Europa left St. John's this morning. Splendid aurora Bay of Bulls to-night, extending over eighty-five degrees of the horizon."

Let any one read this despatch, sentence by sentence, noting the minuteness of the details—which could not be known or conjectured—such as the appearance of yellow fever at New Orleans, with the number of deaths a day; the sailing or arrival of seven steamers; the number of passengers for Fraser's River, etc.—and then examine the London Times, in which all these items appeared Monday morning, August thirtieth, and if he does not admit that collusion or deception is out of the question, no amount of evidence could convince him.

We will give but one proof more. On the last day of August, the day before the cable ceased to work, Valentia sent to Newfoundland two messages for the British Government, both signed by "the Military Secretary to the Commander-in-Chief, Horse Guards, London," and addressed—the first to General Trollope, Halifax, which said, "The Sixty-second regiment is not to return to England;" and the other to the General Officer commanding at Montreal, saying: "The Thirty-ninth regiment is not to return to England." The year before (1857) had witnessed the Sepoy Mutiny, which threatened the overthrow of the British Empire in India. The fighting was over, but the country was still agitated, and the Home Government in fear that the rebellion might be renewed, so that it continued to send forward fresh troops. It had sent out orders by mail for these two regiments to embark immediately for home, to be sent to India. But the mutiny being nearly suppressed, this was found not to be necessary, and the prompt countermanding of the order by telegraph saved the British Government, in the cost of transportation of troops, not less than fifty thousand pounds. The despatch to Halifax was received the same day that it was sent from London.

The sending of this despatch, and its almost immediate reception, is attested by an official letter from the War Office in London.

This array of proofs of what took place a quarter of a century ago, may seem superfluous now that experience has made despatches from the other side of the ocean one of the familiar things of our daily life. And yet at that date the achievement was so stupendous, and, as some thought, in its very nature so incredible, that men of the greatest intelligence could not be convinced. The late Mr. Charles O'Conor continued for years to quote the fact that some men believed that a message had actually passed across the Atlantic as the most amazing illustration of human credulity! Happily he lived to see and to appreciate to its full value this latest miracle of scientific discovery, applied by human genius and skill.

CHAPTER XIII.

CAST DOWN, BUT NOT IN DESPAIR.

It takes a long time to recover from a great disaster. When at last the friends of the Atlantic Telegraph were obliged to confess that the cable had ceased to work; when all the efforts of the electricians failed to draw more than a few faint whispers, a dying gasp, from the depths of the sea, there ensued in the public mind a feeling of profound discouragement. For a time this paralyzed all effort to revive the Company and to renew the enterprise. And yet the feeling, though natural, was extreme. If they had not done all they attempted, they had accomplished much. They had at least demonstrated the possibility of laying a cable across the Atlantic Ocean, and of sending messages through it. This alone was no small triumph. So men reasoned when sober reflection returned, and at length the tide of public confidence, which had ebbed so strongly, began to reflow, and once more to creep up the shores of England.

But when a great enterprise has been overthrown, and lies prostrate on the earth, the first impulse of its friends is to call on Cæsar for help. So the first appeal of the Atlantic Telegraph Company was to the British Government. It was claimed, and with reason, that the work was too great to be undertaken by private capital alone. It was a matter, not of private speculation, but of public and national concern. It was, therefore, an object which might justly be undertaken by a powerful government, in the interest of science and of civilization.

To raise capital for a new cable, it was necessary to have some better security than the hazards of a vast and doubtful undertaking. Hence the Company asked the Government to guarantee the interest on a certain amount of stock, even if the second attempt should not prove a success. With such a guarantee, the capital could be raised in London in a day.

In this application they might have been successful, but for an untoward event, which dampened the confidence of the public in all submarine enterprises—the failure of the Red Sea Telegraph. The British Government, anxious to forward communication with India, had given that Company an unconditional guarantee, on the strength of which the capital was raised, and the cable manufactured and laid. But in a short time it ceased to work, a loss which the treasury of Great Britain had to make good. To the public, which did not understand the cause of the failure to be the imperfect construction of the cable, the effect was to impair confidence in all long submarine telegraphs.

Of course, after such an experience, the Government was not disposed to bind itself by such pledges again. It was, however, ready to aid the enterprise by any safe means. It therefore increased its subsidy from fourteen thousand pounds to twenty thousand pounds; and guaranteed eight per cent on six hundred thousand pounds of new capital for twenty-five years, with only one condition—*that the cable should work*. This was a liberal grant, and under the circumstances, was all that could be expected.

Still further to encourage the undertaking, it ordered new soundings to be taken off the coast of Ireland. These were made by Captain Hoskins, of the Royal Navy, and dispelled the fears which had been entertained of a submarine mountain, which would prove an impassable barrier in the path of an ocean telegraph.

But the greatest service which the British Government rendered, was in the long course of experiments which it now ordered, to determine all the difficult problems of submarine telegraphy. In 1859, the year after the failure of the first Atlantic cable, the Board of Trade appointed a committee of the most eminent scientific and engineering authorities in Great Britain to investigate the whole subject. This was composed of Captain Douglas Galton, of the Royal Engineers, then of the War Office, who represented the Government; Professor Wheatstone, the celebrated electrician; William Fairbairn, President of the British Association for the Advancement of Science; George Parker Bidder, whose name ranks with those of Stephenson and Brunel; C. F. Varley, who, in the practical working of telegraphs, had no superior in England; Latimer Clark and Edwin Clark, both engineers, who had had great experience in the business of telegraphing; and George Saward, the Secretary of the Atlantic Telegraph Company.

This Committee sat for nearly two years, at the end of which it made a report to the Government, which fills a very large volume, in which are detailed an immense number of experiments, touching the form and size of cables, their relative strength and flexibility, the power of telegraphing at long distances, the speed at which messages could be sent; and in fine, every possible question, either as to the electrical or engineering difficulties to be overcome. The result of these manifold and laborious experiments is summed up in the following certificate, signed by all who had taken part in this memorable investigation:

"LONDON, 13th July, 1863.

"We, the undersigned, members of the Committee, who were appointed by the Board of Trade, in 1859, to investigate the question of submarine telegraphy, and whose investigation continued from that time to April, 1861, do hereby state, as the result of our deliberations, that a well-insulated cable, properly protected, of suitable specific gravity, made with care, and tested under water throughout its progress with the best known apparatus, and paid into the ocean with the most improved machinery, possesses every prospect of not only being successfully laid in the first instance, but may reasonably be relied upon to continue for many years in an efficient state for the transmission of signals.

> Douglas Galton,   Cromwell F. Varley,
> C. Wheatstone,   Latimer Clark,
> Wm. Fairbairn,   Edwin Clark,
> Geo. P. Bidder,   Geo. Saward."

Thus the years which followed the failure of 1858—though they saw no attempt to lay another ocean cable—were not years of idleness. They were rather years of experiment and of preparation, clearing the way for new efforts and final victory. The Atlantic Telegraph itself had been a grand experiment. It had taught many important truths which could be learned in no other way. Not only had it demonstrated the possibility of telegraphing from continent to continent, but it had been useful even in exposing its own defects, as it taught how to avoid them in the future.

For example, in working the first cable, the electricians had thought it necessary to use a very strong battery. They did not suppose they could reach across the whole breadth of the Atlantic, and touch the Western hemisphere, unless they sent an electric current that was almost like a stroke of lightning; and that, in fact, endangered the safety of the conducting wire. But they soon found that this was unnecessary. God was not in the whirlwind, but in the still, small voice. A soft touch could send a thrill along that iron nerve. It seemed as if the deep were a vast whispering gallery, and that a gentle voice murmured in the ocean caves, like a whisper in a sea-shell, might be caught, so wonderful are the harmonies of nature, by listening ears on remote continents; a miracle of science, that could give a literal meaning to Milton's

> " Airy tongues, that syllable men's names
> On sands, and shores, and desert wildernesses."

These were also years of great progress, not only in the science of submarine telegraphy, but in the construction of deep-sea cables. In spite of the failure of that in the Red Sea, one was laid down in the Mediterranean, 1,535 miles long, from Malta to Alexandria, and another in the Persian Gulf, 1,400 miles long, by which telegraphic communication was finally opened from England to India. Others were laid in different seas and oceans in distant parts of the world. These great triumphs, following the scientific experiments which had been made, revived public confidence, and prepared the way for a fresh attempt to pass the Atlantic.

Yet not much was done to renew the enterprise until 1862. Mr. Field had been indefatigable in his efforts to reanimate the Company. He was continu-

CAST DOWN, BUT NOT IN DESPAIR. 235

ally going back and forth to the British Provinces and to England, urging it wherever his voice could be heard. Yet times were adverse. The United States had been suddenly involved in a tremendous war, which called into the field hundreds of thousands of men, and entailed a burden of many hundreds of millions. While engaged in this life-and-death struggle, and rolling up a mountain of debt, our people had little thought to bestow on great enterprises by land or sea.

And yet one incident of the war forcibly recalled public attention to the necessity of some speedier communication with Europe than by steam. The unhappy Trent affair aroused an angry feeling in Great Britain which nearly resulted in hostilities, all of which might have been prevented by a single word of explanation. As The Times said truly: "We nearly went to war with America because we had not a telegraph across the Atlantic." After such a warning, it was natural that both countries should begin to think seriously of the means of preventing future misunderstanding. Mr. Field went to Washington, and found great readiness on the part of the President and his Cabinet to encourage the enterprise. Mr. Seward wrote to our Minister in London that the American Government would be happy to join with that of Great Britain in promoting this international work. With this encouragement, Mr. Field went to England to urge the Company to renew the undertaking. While in London, he endeavored to obtain from some responsible parties an offer to construct and lay down a cable. Messrs. Glass, Elliot & Co., replied, declaring their willingness to undertake the work, without at first naming the precise terms. They wrote to him under date of February seventeenth:

"SIR: In reply to your inquiries, we beg to state that we should not be willing to manufacture and lay a Submarine Telegraph Cable across the Atlantic, from Ireland to Newfoundland, assuming the entire risk, as we consider that would be too great a responsibility for any single firm to undertake; but we are so confident that these points can be connected by a good and durable cable, that we are willing to contract to do the work, and stake a large sum upon its successful laying and working.

"We shall be prepared in a few days, as soon as we can get the necessary information in regard to what price we can charter suitable ships for the service, to make you a definite offer."

Although it is anticipating a few months in time, we may give here the "definite offer," which was obtained by Mr. Field, on his return to England in the autumn:

"LONDON, October 20, 1862.
"*Cyrus W. Field, Esq., Atlantic Telegraph Company:*

"DEAR SIR: In reply to your inquiries, we beg to state, that we are perfectly confident that a good and durable Submarine Cable can be laid from Ireland to Newfoundland,

and are willing to undertake the contract upon the following conditions:

"First. That we shall be paid each week our actual disbursements for labor and material.

"Second. That when the cable is laid and in working order, we shall receive for our time, services, and profit twenty per cent on the actual cost of the line, in shares of the Company, deliverable to us, in twelve equal monthly instalments, at the end of each successive month whereat the cable shall be found in working order.

"We are so confident that this enterprise can be successfully carried out, that we will make a cash subscription for a sum of twenty-five thousand pounds sterling in the ordinary capital of the Company, and pay the calls on the same when made by the Company.

"Annexed we beg to hand you, for your guidance, a list of all the submarine telegraph cables manufactured and laid by our firm since we commenced this branch of our business, the whole mileage of which, with the exception of the short one between Liverpool and Holyhead, which has been taken up, is at this time in perfect and successful working order. The cable that we had the honor to contract for and lay down for the French Government, connecting France with Algeria, is submerged in water of nearly equal depths to any we should have to encounter between Ireland and Newfoundland.

"You will permit us to suggest that the shore ends of the Atlantic Cable should be composed of very heavy wires, as from our experience the only accidents that have arisen to any of the cables that we have laid have been caused by ships' anchors, and none of those laid out of anchorage ground have ever cost one shilling for repairs.

"The cable that we would suggest for the Atlantic will be an improvement on all those yet manufactured, and we firmly believe will be imperishable when once laid.

"We remain, sir, yours faithfully,

"GLASS, ELLIOT & Co."

The summer of this year Mr. Field spent in America, where he applied himself vigorously to raise capital for the new enterprise. To this end he visited Boston, Providence, Philadelphia, Albany, and Buffalo—to address meetings of merchants and others. He used to amuse us with the account of his visit to the first city, where he was honored with the attendance of a large array of "the solid men of Boston," who listened with an attention that was most flattering to the pride of the speaker, addressing such an assemblage in the capital of his native State. There was no mistaking the interest they felt in the subject. They went still farther, they passed a series of resolutions, in which they applauded the projected telegraph across the ocean as one of the grandest enterprises ever undertaken by man, which they proudly commended to the confidence and support of the American public, after which they went home, feeling that they had done the generous thing in bestowing upon it such a mark of their approbation. *But not a man subscribed a dollar!* Yet it is not necessary to charge them with meanness or hypocrisy. No doubt they felt just what they said. They could not but admire the courage of their countryman. It was inspiring to hear him talk. Yet these solid men were never lifted off their feet so far as to forget the main chance. What were to be the returns for this magnificent adventure? Peering into the future, the prospect of dividends was very remote. In fact they looked upon the Atlantic Telegraph as a sort of South Sea Bubble, an airy fancy, which would go up like a balloon, never to return to earth again. So, like the high priest and the Levite, they passed by on the other side.

Other cities were equally gracious, equally complimentary, but equally prudent. In New York he succeeded better, but only by indefatigable exertions. He addressed the Chamber of Commerce, the Board of Brokers, and the Corn Exchange, and then he went almost literally from door to door, calling on merchants and bankers to enlist their aid. The result was, subscriptions amounting to about seventy thousand pounds, the whole of which was due to persevering personal solicitation. Even of those who subscribed, a large part did so more from sympathy and admiration of his indomitable spirit than from confidence in the success of the enterprise.

In England, however, the subject was better understood. For obvious reasons, the science of submarine telegraphy had made greater advances in that country than in ours. As England is an island, she is obliged to hold all her telegraphic communication with the continent by cables under the sea. She has colonial possessions in all parts of the world. A power that rules so large a part of the earth cannot be shut up in her island home. No one has depicted the extent of her dominion in nobler phrase than our own Webster when he speaks of the imperial sway which "has dotted the face of the whole globe with its possessions and military posts, whose morning drumbeat, following the sun and keeping company with the hours, encircles the whole earth with one continuous and unbroken strain of the martial airs of England." Was it strange that this mother of nations should reach out her long arms to embrace her distant children?

Hence it was that the subject of submarine telegraphs was so much better understood in England than in America, not only by scientific men, but by capitalists. The appeal could be made to them with more assurance of intelligent sympathy. And yet so vast was the undertaking, that it required ceaseless effort to roll the stone to the top of the mountain, and the result was not completely achieved till the beginning of the year 1864.

CHAPTER XIV.

THE EXPEDITION OF 1865.

It is a long night which has no morning. At last the day is breaking. While weary eyes are watching the East, daylight comes over the sea. Five years have passed away, and though the time seemed long as an Arctic winter, that only made more bright the rising of the sun. Those years of patient experiment, when scientific men were applying tests without number, and submarine lines were feeling their way along the deep-sea floor in all the waters of the world, at last brought forth their fruit in that renewed confidence which is the forerunner of victory.

So strong was this feeling, that as early as August, 1863, although the capital was not raised, the Board advertised for proposals for a cable suitable to be laid across the Atlantic Ocean; and in order to leave invention entirely unfettered, abstained from any dictation as to the form or materials to be adopted, merely stipulating for a working speed of eight words a minute.

To this request they received, in the course of a few weeks, seventeen different proposals from as many companies, many of them firms of large wealth and experience. These different tenders, with the numerous specimens of cable and materials, were at once submitted to a Consulting Committee composed in part of members of the Committee which had already rendered such service by its advice. It consisted of Captain Douglas Galton, William Fairbairn, Professor C. Wheatstone, William Whitworth and Professor William Thomson. There were no more distinguished engineers and electricians in the world. They examined all the proposals, and then, taking up one by one the different samples of cable, caused them in turn to be subjected to the severest tests. This took a long time, as it required a great number of experiments; but the result was highly satisfactory. The Committee were all of one mind, and recommended unanimously that the Board should accept the tender of Messrs. Glass, Elliot & Co., and the general principle of their proposed cable; but advised that before settling the final specification, every portion of the material to be employed should be tested with the greatest care, both separately and in combination, so as to ascertain what further improvements could be made. To this the manufacturers readily consented, feeling a noble ambition to justify the confidence of the Committee and the public. They provided abundant materials for fresh experiments. New cables were made and tested in different lengths; and experiments

THE EXPEDITION OF 1865.

were also tried upon different qualities of wire and hemp, that were to compose its external protection. The result of all these investigations was the selection of a model which seemed to combine every excellence, and to approach absolute perfection.

Such was the cable which this eminent firm offered to manufacture, and to lay across the Atlantic, and that on terms so favorable, that it seemed as if it could not be difficult to raise the capital and proceed with the work. Indeed, a contract was partially made to that effect. So confident was Mr. Field, who was then in London, that an expedition would sail the following summer, that he insured his stock, part of it only against ordinary sea-risks, but part also to be laid and to work! But hardly had he left England before there was some unforeseen hitch in the arrangements, the money was not forthcoming, or some of the conditions were not complied with, and he had the mortification to receive letters, saying that the whole enterprise was postponed for another year!

This was indeed discouraging. Yet this sudden dropping of the scheme did not imply a loss of interest or of faith on the part of those embarked in it. They believed in it as much as ever. But the general public did not respond to the call for more capital. Alas that the noblest enterprises should so often be delayed or defeated by the want of money! Capital is always cautious and timid, and follows slowly in the path of great discoveries. If Columbus, instead of the patronage of a Queen full of womanly enthusiasm, had depended on a stock company for the means for his expedition, he might never have sailed from the shores of Spain. Happy was it for mankind that his faith and patience did not wear out, while going from court to court, and kingdom to kingdom, and almost begging his way from door to door!

But it is not in human nature—least of all in American nature—to despond long. Though ten years of constant defeat would seem to have wrought a lasting discouragement, yet again and again did the baffled spirit of enterprise return to the attempt. In January, 1864, Mr. Field was once more on his way to England. He found the Directors, as before, deeply interested in the enterprise, and wishing it success. With a grateful heart he bore witness to their unfaltering courage. But mere courage and good wishes would not lay the Atlantic Telegraph. Yet what could they do? They could not be expected to advance all the capital themselves. They had already subscribed liberally, and he could not ask them to do more. But with all the efforts that had been made in England and America, not half the capital was yet raised. The machinery was in a dead lock, with little prospect of being able to move. It was the misfortune of the enterprise that there was no one man who made it his sole and exclu-

THE EXPEDITION OF 1865.

sive charge. The Board of Directors contained some of the best men in London. But they were, almost without exception, engaged in very large affairs of their own, with no leisure to make a public enterprise their special care. To insure success, it needed a trial of the one-man power—one brain, planning night and day; one agency incessantly at work, stirring up· directors, contractors, and engineers; and one will pushing it forward by main strength. This was the force now to be applied.

The first element needed to put life into the old system was an infusion of new blood—new capital and new men. While the enterprise was in this state of collapse, Mr. Field addressed himself to a gentleman with whom, until then, he had no personal acquaintance, but who was well known in London as one of the largest capitalists of Great Britain—Mr. Thomas Brassey. Their first interview was somewhat remarkable. Referring to it a few months after, Mr. Field said:

"When I arrived in this country, in January last, the Atlantic Telegraph Company trembled in the balance. We were in want of funds, and were in negotiations with the government, and making great exertions to raise the money. At this juncture I was introduced to a gentleman of great integrity and enterprise, who is well known, not only for his wealth, but for his foresight, and in attempting to enlist him in our cause he put me through such a cross-examination as I had never before experienced. I thought I was in the witness-box. He inquired of me the practicability of the scheme —what it would pay, and every thing else connected with it; but before I left him, I had the pleasure of hearing him say that it was a great national enterprise that ought to be carried out, and, he added, I will be one of ten to find the money required for it. From that day to this he has never hesitated about it, and when I mention his name, you will know him as a man whose word is as good as his bond, and as for his bond, there is no better in England."

Having thus secured one powerful ally, Mr. Field took courage in the hope to find another. He says:

"The words spoken by Mr. Brassey in the latter part of January, 'Let the Electric Telegraph be laid between England and America,' encouraged us all, and made us believe we should succeed in raising the necessary capital, and I then went to work to find nine other Thomas Brasseys (I did not know whether he was an Englishman, a Scotchman, or an Irishman; but I made up my mind that he combined all the good qualities of every one of them), and after considerable search I met with a rich friend from Manchester, Mr. [now Sir] John Pender, and I asked him if he would second Mr. Brassey, and walked with him from 28 Pall Mall to the House of Commons, of which he is a member. Before we reached the House, he expressed his willingness to do so to an equal amount."

This was putting strong arms to the wheel. A few days after, a combination was formed to carry on the whole business of making Submarine Telegraphs, by a union of the Gutta-Percha Company with the firm of Glass, Elliot & Co., the principal manufacturers of sea cables, making one grand concern, to be called The Telegraph Construction and Maintenance Company. These two great capitalists entered into the new organization, of which Mr. Pender was made Chairman. The Gutta-Percha Company brought in still further strength to the joint enterprise, in the person of Mr. Willoughby Smith, their electrician, and of Mr. John Chatterton, the inventor of the insulating material known as Chatterton's Compound. The union of all these men made a combination of practical skill and financial ability, such as could be found in few companies in England or in the world. Mr. R. A. Glass was chosen Managing Director—a gentleman who seemed born to be a manager, such power had he of gathering about him talent in every department and combining all into one organization. Reënforced by such powerful aid, the new Company now came forward, and offered at one stroke to take all the remaining stock of the Company. This was more than half the whole capital. As yet, of the £600,000 required, but £285,000 had been subscribed. Now this princely Company offered to take the balance themselves— £315,000. They did more, they took £100,000 of bonds: and so by one dead lift these stalwart Englishmen took the whole enterprise on their broad shoulders. From that hour the problem was solved. Thus after a dead lock of six months the wheels were unloosed, and the gigantic machinery began to revolve.

This was a triumph worthy to be honored in the way that Englishmen love, by a little festivity; and as it chanced to be now ten years since Mr. Field had embarked in the enterprise, the pleasant thought occurred to him of getting his friends together to celebrate the anniversary. Accordingly, on the fifteenth of March, he invited them to dine together at the Buckingham Palace Hotel. It was a joyous occasion, and called forth the usual amount of toasts and speeches. Of the latter, those of Mr. Adams, the American Minister, and of John Bright, were widely copied in the United States. The next day was the annual meeting of the Atlantic Telegraph Company, when the Chairman, the Right Hon. James Stuart Wortley, thus referred to the gathering of the night before:

"Without saying any thing to detract from my deep gratitude to the other Directors, I cannot help especially alluding to Mr. Cyrus Field, who is present to-day, and who has crossed the Atlantic thirty-one times in the service of this Company, having celebrated at his table yesterday the anniversary of the tenth year of the day when he first left Boston in the service of the Company. Collected round his table last night was a company of distinguished men—members of Parliament, great capitalists, distinguished merchants and manufacturers, engineers and men of science, such as is rarely found together even in the highest house in this great metropolis. It was very agreeable to see an American citizen so surrounded. It was still more gratifying, inasmuch as we were there to celebrate the approaching accomplishment of the Atlantic Telegraph."

This was a congratulation on an escape from death, for their cherished scheme had just passed through a critical period of its history. The enterprise had been in great danger of abandonment—at least for years, a peril from which it had been rescued only by the most prompt and vigorous effort.

Thus after infinite toil, the wreck of old disasters was cleared away, and the mighty task begun anew. The works of the Telegraph Construction and Maintenance Company were the largest in the world, and all their resources were now put in requisition. Never did greater care preside over a public enterprise. It was a case in which the motive of interest was seconded or overborne by pride and ambition. A cable was to be made to span the Atlantic Ocean, and to join the hemispheres; and they were determined to produce a work that should be as nearly perfect as human skill could make it. The Scientific Committee, that had so long investigated the subject, had approved a particular form of cable, as "the one most calculated to insure success in the present state of our experimental knowledge respecting deep-sea cables," but at the same time recommended the utmost vigilance at every stage of the manufacture. These precautions deserve to be noted, as showing with what jealous care science watches over the birth of a great enterprise, and prescribes the conditions of success. They recommended:

That the conductivity of the wire should be fixed at a high standard, certainly not less than eighty-five per cent; that the cable should be at least equal to the best ever made; that the core should be electrically perfect; that it should be

OLD ATLANTIC CABLE, 1858.

NEW ATLANTIC CABLE, 1865.

tested under hydraulic pressure, and at the highest pressure attainable in the tanks at the Company's works; that after this pressure, the core should be examined again, and before receiving its outer covering, be required to pass the full electrical test under water; that careful and frequent mechanical tests be made upon the iron wire and hemp as to their strength; that special care be given to the joints, where different lengths of cable were spliced together; and that when com-

THE EXPEDITION OF 1865.

pleted, the whole be tested under water for some length of time, at a temperature of seventy-five degrees.

This was higher by forty degrees than the temperature of the Atlantic. The insulation is improved by cold; so that, if it remained perfect in this warm water, it could not fail in the icy depths of the ocean.

After passing through such elaborate tests, all will be glad to see the final product of so much care and skill. As the long line begins to reel off from the great wheels and drums, we may examine it in its completed and more perfect form. It is only necessary to compare it with the cable laid in 1858, to show its immense superiority. A glance at the two as they appear on the preceding page will show that the cable had *grown* since first it was planted in the ocean, as if it were a living product of the sea. This growth had been in every part, from core to circumference.

First, the central copper wire, which was the spinal cord, the nerve along which the centre current was to run, was nearly three times as large as before. Prof. Thomson had long seen that this was a condition of success. While joining heartily in the attempts of 1857–58, he felt that an error was committed in the smallness of the cable; that the copper conductors and the gutta-percha covering should both be much larger. The old conductor was a strand consisting of seven fine wires, six laid round one, and weighed only one hundred and seven pounds to the mile. The new was composed of the same number of wires, but weighed three hundred pounds to the mile. As it was made of the finest copper that could be obtained in the world, it was a perfect conductor. Next, to secure insulation, it was first imbedded for solidity in Chatterton's compound, a preparation impervious to water, and then covered with four layers of gutta-percha, which were laid on alternately with four thin layers of Chatterton's compound. The old cable had but three coatings of gutta-percha, with nothing between. Its entire insulation weighed but two hundred and sixty-one pounds to the mile, while that of the new weighed four hundred pounds.

But a conductor ever so perfect, with insulation complete, was useless without proper external protection, to guard it against the dangers which must attend the long and difficult process of laying it across the ocean. The old cable had broken a number of times. The new must be made stronger. To this end it was incased with ten solid wires of the best iron, or rather, of a soft steel, like that used in the making of Whitworth's cannon. This made the cable much heavier than before. The old cable weighed but twenty cwt. to the mile, while the new one reached thirty-five cwt. and three quarters. But mere size and weight were nothing, except as they indicated increased strength. This was secured, not only by the larger iron wires, but by a further coating of rope.

Each wire was surrounded separately with five strands of Manilla yarn, saturated with a preservative compound, and the whole laid spirally round the core, which latter was padded with ordinary hemp, saturated with the same preservative mixture. This rope covering was important in several respects. It kept the wires from coming in contact with the salt water, by which they might be corroded; and while it added greatly to the strength of the cable, it gave it also its own flexibility—so that while it had the strength of an iron chain, it had also the lightness and flexibility of a common ship's rope. This union of two qualities was all-important. The great problem had been to combine strength with flexibility. Mere dead weight was an objection. The new cable, though nearly twice as heavy as the old in air, when immersed in water, weighed but a trifle more; so that it was really much lighter in proportion to its size. This increased lightness was a very important matter in laying the cable, as it caused it to sink slowly. The old cable, though smaller, was heavy almost as a rod of iron, so that, as it ran out, it dropped at an angle which exposed it to great danger in case of a sudden lurch of the ship. Thus in 1857 it was broken by the stern of the Niagara being thrown up on a wave just as the brakes were shut down. Now the cable, being partially buoyed by the rope, would float out to a great distance from the ship, and sink down slowly in the deep waters.

By this combination of rope and iron, a cable was secured two and a half times as strong as the old—the breaking strain of the former having been three tons five cwt., and of the latter seven tons and fifteen cwt. Or, to put it in another form, the contract strain of the former was less than five times its own weight per mile in water; so that if the cable had been laid in some parts of the Atlantic, where the ocean is more than five miles deep, it would have broken under the enormous strain. But the contract strain of the new cable was equal to *eleven* times its weight per mile in water, which, as the greatest depth of water to be passed was but two and a half miles, rendered the cable more than four times as strong as was required.

This great chain which was to bind the sea was to be 2,300 nautical miles long, or nearly 2,700 statute miles! But where could this enormous bulk be stowed? Its weight would sink the Spanish Armada. In 1858, the cable loaded down two of the largest ships of war in the world, the Niagara and the Agamemnon. Yet now one much larger and bulkier was to be taken on board. This might have proved a serious embarrassment, but that a few years before there had been built in England a ship of enormous proportions. The Great Eastern, whose iron walls had been reared by the genius of Brunel, had been for ten years waiting for "a mission." As a specimen of marine architecture she was perfect. She walked the

waters in towering pride, scarce bending her imperial head to the waves that broke against her sides, as against the rocks of the shore. But with all her noble qualities, she was too great for the ordinary demands of commerce. Her very size was against her; and while smaller ships, on which she looked down with contempt, were continually flying to and fro across the sea, this leviathan,
> Hugest of all God's works
> That swim the ocean stream,

could find nothing worthy of her greatness. Here then was the vessel to receive the Atlantic cable.

Seeing her fitness for the purpose, a few of the gentlemen who were active in reviving the Atlantic Telegraph combined to purchase her, as she was about to be sold. One of them went down with all speed to Liverpool, and the next day telegraphed that the big ship was theirs. The new owners at once put her at the service of the Atlantic Company, with the express agreement that any compensation for her use should depend on the success of the expedition.

Next to the good fortune of finding such a ship ready to their hands, was that of finding an officer worthy to command her. Captain James Anderson, of the China, one of the Cunard steamers, had long been known to the travelling public, both of England and America, and no one ever crossed the sea with him without the strongest feeling of respect for his manly and seamanly qualities. A thorough master of his profession, having followed the sea for a quarter of a century, he was also a man of much general intelligence, and of no small scientific attainments. But it was something more than this which inspired such confidence. It was his ceaseless watchfulness. He always carried with him a feeling of religious responsibility for the lives of all on board, and for every interest committed to him. A man of few words, modest in manner, he was yet clear in judgment and prompt in action. This vigilance was especially marked in moments of danger. When a storm was gathering, all who saw that tall figure on the wheel-house, watching with a keen eye every spar in the ship and every cloud in the horizon, felt a new security from being under his care. Such was the man to be put in charge of a great expedition. He was recommended by Mr. Field in the strongest terms, and was chosen unanimously by the Board. The Cunard Company, with great generosity, consented to give up his services, valuable as they were, to forward an enterprise of such public interest. Being thus free, he accepted the trust, and entered upon it with enthusiasm. How well he fulfilled the expectations of all, the sequel will show.*

* Nearly a year and a half after this, when the cable was safely landed in Newfoundland, Captain Anderson, still on board the Great Eastern, in a letter to a friend, thus referred to his first connection with the Atlantic Telegraph:—

"I cannot tell you how I have felt since our success. It is only sev-

The work now went on with speed. The wheels began to hum, and the great drums to reel off that line which, considering the distance it was to span, was hardly to be measured by miles, but rather by degrees of the earth's surface. Mere figures give but a vague impression of vast spaces. But it is a curious fact, ascertained by an exact computation, that if all the wires of copper and of iron, with the layers that made up the core and the outer covering, and the strands of yarn that were twisted into this one knotted sea-cable, were placed end to end, the whole length would reach from the earth to the moon!

enteen months since I first walked up to the top of the paddle-box of this ship at Sheerness, upon a dark, rainy night—reviewed my past career in my mind, and tried to look into the future, to see what I had undertaken, and realize if possible what this new step would develop. I cannot say I believed much in cables; I rather think I did not; but I did believe Mr. Field was an earnest man, of great force of character, and working under a strong conviction that what he was attempting was thoroughly practicable; and I knew enough of the names with which he had associated himself in the enterprise to feel that it was a real, true, honest effort, worthy of all the energy and application of one's manhood; and come what might in the future, I resolved to do my very utmost, and to do nothing else until it was over. More completely however than my resolve foreshadowed, I dropped inch by inch, or step by step, into the work, until I had no mind, no soul, no sleep, that was not tinged with cable. In a word I accuse Mr. Field of having dragged me into a vortex, that I could not get out of, and did not wish to try—and the sum total of all this is, to lay a thread across an ocean! Dr. Russell compared it to an elephant stretching a cobweb, and there lay its very danger: the more you multiply the mechanism, the more you increase the risk."

As it came from the works in its completed state, it was plunged in water, to make it familiar with the element which was to be its future home. In the yards of the Company stood eight large tanks, which could hold each a hundred and forty miles. Here the cable was coiled to "hibernate," till it should be wanted for use the coming spring.

Seeing the work thus well under way, with no chance of another disastrous check, Mr. Field left England with heart at rest, and returned to America for the winter. But the first days of spring saw him again on the Atlantic. He reached England on the eighteenth of March. His visit was more satisfactory than a year before. The work was now well advanced. It was a goodly sight to go down to Morden Wharf at Greenwich, and see the huge machinery in motion, spinning off the leagues of deep-sea line. The triumph apparently was near at hand. It seemed indeed a predestined thing that the cable should finally be laid in the year of grace 1865—the end for which he had so faithfully toiled since 1858—seven weary years—as long as Jacob served for Rachel! But, less fortunate than Jacob, he was doomed to one more disappointment. At present, however, all looked well, and he could not but regard the prospect with satisfaction.

Having no more drudgery of raising money, he had now a few weeks' leisure to take a voyage up the

THE EXPEDITION OF 1865.

Mediterranean. The canal across the Isthmus of Suez, which had been so long in progress, under the supervision of French engineers, was at length so far advanced that the waters of the Mediterranean were about to mingle with those of the Red Sea, and delegates were invited to be present from all parts of the world. An invitation had been sent to the Chamber of Commerce in New York, and Mr. Field, then starting for Europe, was appointed as its representative. The visit was one of extraordinary interest. The occasion brought together a number of eminent engineers from every country of Europe, in company with whom this stranger from the New World visited the most ancient of kingdoms to see the spirit of modern enterprise invading the land of the Pyramids.

He returned to England about the first of May to find the work nearly completed. The cable was almost done, and a large part of it was already coiled on board the ship. This was an operation of much interest, which deserves to be described. The manufacture had begun on the first of September, and had gone on for eight months without ceasing, the works turning out fourteen miles a day even during the short days of winter. As the spring advanced, and the days grew longer, the amount was of course much increased. But by the last of January they had already accumulated about nine hundred miles of completed cable, when began the long and tedious work of transferring it to the Great Eastern. It was thus slow, because it could not be made directly from the yard to the ship. The depth of water at Greenwich was not such as to allow the Great Eastern to be brought up alongside the wharf. She was lying at Sheerness, thirty miles below, and the cable had to be put on board of lighters and taken down to where she lay in the stream. For this purpose the Admiralty had furnished to the Company two old hulks, the Iris and the Amethyst, which took their loads in turn. When the former had taken on board some two hundred and fifty tons of cable, she was towed down to the side of the Great Eastern, and the other took her place.

This was an operation which could not be done with speed. With all the men who could be employed, they coiled on board only about two miles an hour, or twenty miles a day—at which rate it would take some five months. The work began on the nineteenth of January, early in the morning, and continued till June, before all was safely stowed on board. The Great Eastern herself had been fitted up to receive her enormous burden. It was an object to stow the cable in as few coils as possible. Yet it could not be all piled in one mass. Such a dead weight in the centre of the ship would cause her to roll fearfully. If coiled in one circle, it was computed that it would nearly fill Astley's theatre from the floor of the circus to the roof—making a pile fifty-eight feet wide and sixty feet high. To distribute this enormous bulk and weight, it was disposed in three tanks—one aft, one amidships, and one forward. The latter, from the shape of the ship, was a little smaller than the others, and held only six hundred and thirty-three miles of cable, while the two former held a little over eight hundred each. All were made of thick wrought-iron plates, and water-tight, so that the cable could be kept under water till it was immersed in the sea.

Thus with her spacious chambers prepared for the reception of her guest, the Great Eastern opened her doors to take in the Atlantic cable; and long as it was, and wide and high the space it filled, it found ample verge and room within her capacious sides. Indeed, it was the wonder of all who beheld it, how, like a monster of the sea, she devoured all that other ships could bring. The Iris and the Amethyst came up time after time and disgorged their iron contents. Yet this leviathan swallowed ship-load after ship-load, as if she could never be satisfied. A writer who visited her when the cable was nearly all on board, was at a loss to find it. He looked along the deck, from stem to stern, but not a sign of it appeared. How he searched, and how the wonder grew, he tells in a published letter. After describing his approach to the ship, and climbing up her sides and his survey of her deck, he proceeds:

"But it is time that we should look after what we have mainly come to see, the telegraph cable. To our intense astonishment, we behold it nowhere, although informed that there are nearly two thousand miles of it already on board, and the remaining piece—a piece long enough to stretch from Land's End to John O'Groat's—is in course of shipment. We walk up and down on the deck of the Great Eastern without seeing this gigantic chain which is to bind together the Old and the New World ; and it is only on having the place pointed out to us that we find where the cable lies and by what process it is taken on board. On the side opposite to where we landed, deep below the deck of our giant, there is moored a vessel surmounted by a timber structure resembling a house, and from this vessel the wonderful telegraph cable is drawn silently into the immense womb of the Great Eastern. The work is done so quietly and noiselessly, by means of a small steam-engine, that we scarcely notice it. Indeed, were it not pointed out to us, we would never think that that little iron cord, about an inch in diameter, which is sliding over a few rollers and through a wooden table, is a thing of world-wide fame—a thing which may influence the life of whole nations; nay, which may affect the march of civilization. Following the direction in which the iron rope goes, we now come to the most marvellous sight yet seen on board the Great Eastern. We find ourselves in a little wooden cabin, and look down, over a railing at the side, into an immense cavern below. This cavern is one of the three 'tanks' in which the two-thousand-mile cable is finding a temporary home. The passive agent of electricity comes creeping in here in a beautiful, silent manner, and is deposited in spiral coils, layer upon layer. It is almost dark at the immense depth below, and we can only dimly discern the human figures through whose hands the coil passes to its bed. Suddenly, however, the men begin singing. They intone a low, plaintive song of the sea; something like Kingsley's

> 'Three fishers went sailing away to the West,
> Away to the West as the sun went down—'

the sounds of which rise up from the dark, deep cavern with startling effect, and produce an indescribable impression.

"We proceed on; but the song of the sailors who are taking charge of the Atlantic Telegraph cable is haunting us like a dream. In vain our guide conducts us all over the big ship, through miles of galleries, passages, staircases, and promenades; through gorgeous saloons, full of mirrors, marbles, paintings, and upholstery, made 'regardless of expense;' and through buildings crowded with glittering steam apparatus of gigantic dimensions, where the latent power of coal and water creates the force which propels this monster vessel over the seas. In vain our attention is directed to all these sights; we do not admire them; our imagination is used up. The echo of the sailors' song in the womb of the Great Eastern will not be banished from our mind. It raises visions of the future of the mystic iron coil under our feet—how it will roll forth again from its narrow berth; how it will sink to the bottom of the Atlantic, or hang from mountain to mountain far below the stormy waves; and how two great nations, offsprings of one race and pioneers of civilization, will speak through this wonderful coil, annihilating distance and time. Who can help dreaming here, on the spot where we stand? For it is truly a marvellous romance of civilization, this Great Eastern and this Atlantic Telegraph cable.

Even should our age produce nothing else, it alone would be the triumph of our age."

As the work approached completion, public interest revived in the stupendous undertaking, and crowds of wonder-seekers came down from London to see the preparations for the expedition. Even if not admitted on board, they found a satisfaction in sailing round the great ship, in whose mighty bosom was coiled this huge sea-serpent. It had also many distinguished visitors. Among others, the Prince of Wales came to see the ocean girdle which was to link the British islands with his future dominions beyond the sea.

At length, on the twenty-ninth of May, almost the last day of Spring, the manufacture of the cable was finished. The machines which for eight months had been in a constant whirl, made their last round. The tinkling of a bell announced that the machinery was empty, and the mighty work stood completed. It only remained that it should be got on board, and the ship prepared for her voyage. Hundreds of busy hands were at work without ceasing, and yet it was six weeks before she was ready to put to sea.

It may well be believed that it was no small affair to equip such an expedition. Beside the enormous burden of the cable itself, the Great Eastern had to take on board seven or eight thousand tons of coal, enough for a fleet, to feed her fires. Then she carried about five hundred men, for whom she had to make provision during the weeks they might be at sea. The stores laid in were enough for a small army. Standing on the wheel-house, and looking down, one might fancy himself in some large farm-yard of England. There stood the motherly cow that was to give them milk; and a dozen oxen, and twenty pigs, and a hundred and twenty sheep, while whole flocks of ducks and geese, and fowls of every kind, cackled as in a poultry-yard. Beside all this live stock, hundreds of barrels of provisions, of meats and fruits, were stored in the well-stocked larder below. Thus laden for her voyage, the Great Eastern had in her a weight, including her own machinery, of twenty-one thousand tons—a burden almost as great as could have been carried by the whole fleet with which Nelson fought the battle of Trafalgar.

As the time of departure drew near, public curiosity was excited, and there was an extraordinary desire to witness the approaching attempt. The Company was besieged by applications from all quarters for permission to accompany the expedition. Had these requests been granted, on the scale asked, even the large dimensions of the Great Eastern could hardly have been sufficient for the crowds on board. The demand was most pressing for places for newspaper correspondents. These came not only from England, but from France and America. Almost every journal in London claimed the privilege of being represented. The result was what might have been expected. As it was impossible to satisfy all, and to discriminate in favor of some, and exclude others, would seem partial and unjust, they were finally obliged to exclude all. Of course this gave great offence. There was an outcry in England and in the United States at what was denounced as a selfish and suicidal policy. But it is doubtful whether any other possible course would have given better satisfaction.

Whether the Managers erred in this or not, it should be said that they applied the same inexorable rule to themselves—even Directors of the Company being excluded, unless they had some special business on board.

It should be borne in mind that the expedition was not under the control of the Atlantic Telegraph Company at all, but of the Telegraph Construction and Maintenance Company, which had undertaken the work in fulfilment of a contract with the former Company to manufacture and lay down a cable across the Atlantic, in which it assumed the whole responsibility, not only making the cable, but chartering the ship and appointing the officers, and sending its own engineers to lay it down. Of course it had an enormous stake in the result. Hence it felt, not only authorized, but bound, to organize the expedition solely with reference to success. It was not a voyage of pleasure, but for business; for the accomplishment of a great and most difficult undertaking. Hence it was right that

THE EXPEDITION OF 1865.

the most strict rules should be adopted. Accordingly there was not a man on board who had not some business there. As the voyage promised to be one of the utmost practical interest to electricians and engineers, several young men were received as assistants in the testing-room or in the engineers' department; but there was no person who was not in some way engaged on the business of one or the other company, or connected with the management of the ship. Except Mr. Field, not an Atlantic Telegraph Director accompanied the expedition; and he represented also the Newfoundland Company. Mr. Gooch, M.P., was at once a Director of the Telegraph Construction and Maintenance Company, and Chairman of the Board that owned the Great Eastern, and so represented both those companies which had so great a stake in the result.

Thus the whole business was in the hands of the Telegraph Construction and Maintenance Company. It had its own officers to man the expedition—the captain and crew to sail the ship—its engineers to lay the cable—and its electricians to test it. Even the eminent electricians, Professor Thomson and Mr. Varley, who were on board in the service of the Atlantic Telegraph Company, were not allowed to interfere, *nor even to give advice* unless it were asked for in writing, and then it was to be given in writing. Their office was only to test the cable when laid, to pass messages through it from Newfoundland to Ireland, and to report it complete.

So rigorous were the rules which governed this memorable voyage. The whole enterprise was organized as completely as a naval expedition. Every man had his place. As when a ship is going into battle, everybody is sent below that has not some business on deck, so it is not strange that in such a critical enterprise they did not want a host of supernumeraries on board.

Yet the Company was not unmindful of the anxiety of the public for news, and since it could not give a place to many correspondents, it engaged one, and that the best—W. H. Russell, LL.D., the well-known correspondent of the London Times in the Crimea and in India. This brilliant writer was engaged to accompany the expedition—not to praise without discrimination, but to report events faithfully from day to day. He was accompanied by two artists, Mr. O'Neill and Mr. Dudley, to illustrate the scenes of the voyage. Thus the Company made every provision to furnish information and even entertainment to the public. Several of these gentlemen afterward wrote accounts for different magazines—Blackwood, Cornhill, and Macmillan's. Their different reports, and especially the volume of Dr. Russell, which combines the accuracy and minuteness of a diary kept from day to day, with brilliant descriptions, set off by illustrations

THE EXPEDITION OF 1865. 269

from drawings of Mr. Dudley, furnish the public as full and complete an account as if there had been a special correspondent for every journal of England and America.

But if the public at large were very properly excluded, the organization on board was perfect and complete. At the head was Captain Anderson, of whom we have already spoken. As his duties would be manifold and increasing, he had requested the aid of an assistant commander, and Captain Moriarty, R. N., who had been in the Agamemnon in 1858, was permitted by the Admiralty to accompany the ship, and to give the invaluable aid of his experience and skill. The government also generously granted two ships of war, the Sphinx and the Terrible, to attend the Great Eastern. Thus the whole equipment of the expedition was English. Of the five hundred men on board the Great Eastern, there was but one American, and that was Mr. Field.

The engineering department was under charge of Mr. (now Sir) Samuel Canning, who, as the representative of the Telegraph Construction and Maintenance Company, was chief in command of all matters relating to laying the cable. For this responsible position no better man could have been chosen. Before the voyage was ended, he had ample opportunity to show his resources. He was ably seconded by Mr. Henry Clifford. Both these gentlemen had been on board the

Agamemnon in the two expeditions of 1858. They had since had large experience in laying submarine cables in the Mediterranean and other seas. It was chiefly by their united skill that the paying-out machinery had been brought to such perfection, that throughout the voyage it worked without a single hitch or jar. They had an invaluable helper in Mr. John Temple.

The electrical department was under charge of Mr. De Sauty, who had had long experience in submarine telegraphs, and who was aided by an efficient corps of assistants. Professor Thomson and Mr. Varley, as we have said, were there to examine and report for the Atlantic Company. All these gentlemen had been unceasing in their tests of the cable in every form, both while in the process of manufacture and after it was coiled in the Great Eastern. The result of their repeated tests was to demonstrate that the cable was *many times more perfect than the contract required.* With such marvellous delicacy did they test the current of electricity sent through it, that it was determined that of one thousand parts, over nine hundred and ninety-nine came out at the other end!

To complete this organization and equipment caused such delays as excited the impatience of all on board. But at length, when midsummer had fully come—at noon of Saturday, July fifteenth—the song of the sailors sounded the *chant du départ.* The Great Eastern was then lying at the Nore, and she seemed to cling to the English soil which she had griped with a huge Trotman weighing seven tons, held fast by a chain whereof every link weighed seventy pounds! To wrench this ponderous anchor from its bed required the united strength of near two hundred men. At last the bottom lets go its hold, the anchor swings to the bow, the gun is fired, and the voyage is begun. A fleet of yachts and boats raise their cheers as the mighty hull begins to move. But mark how carefully she feels her way, following the lead of yonder little steamer, the Porcupine, the same faithful guide that seven years before led the Niagara up Trinity Bay one night when the faint light of stars twinkled on all the surrounding hills. Slowly they near the sea. Now the cliffs of Dover are in sight, and bidding her escort adieu, the Great Eastern glides along by the beautiful Isle of Wight, and then quickening her speed, with a royal sweep, she moves down the Channel. Off Falmouth she picked up the Caroline, a small steamer, which had left several days before with the shore end on board. She was laboring heavily with her burden, and made little headway in the rough waves. But the Great Eastern took her in tow, and she followed like a ship's boat in the wake of the monarch of the seas.

Thus they passed round to the coast of Ireland, to that Valentia Bay where, eight years before, the Earl of Carlisle gave his benediction on the departure of the Niagara and the Agamemnon, and where, a year later, the gallant English ship brought her end of the cable safely to the shore.

The point of landing had been changed from Valentia harbor five or six miles to Foilhommerum Bay, a wild spot where huge cliffs hang over the waves that here come rolling in from the Atlantic. On the top, an old tower of the time of Cromwell tells of the bloody days of England's great civil war. It is now but a mossy ruin. Here the peasants who flocked in from the country pitched their booths on the green sward, and looked down from the dizzy heights on the boats dancing in the bay below. At the foot of the cliff, a soft, sandy beach forms a bed for the cable, and here, as it issues from the sea, it is led up a channel which had been cut for it in the rocks.

As the shore end was very massive and unwieldy, it could not be laid except in good weather; and as the sea was now rough, the Great Eastern withdrew to Bantry Bay, to be out of the way of the storms which sometimes break with fury on this rock-bound coast.

On Saturday this preliminary work was completed, the heavy shore end was carried from the deck of the Caroline across a bridge of boats to the beach, and hauled up the cliffs amid the shouts of the people. When once it was made fast to the rocks, the little steamer began to move, and the huge coil slowly unwound, and like a giant awakened, stretched out its long iron arms. By half-past ten o'clock at night the hold was empty, the whole twenty-seven miles having been safely laid, and the end buoyed in seventy-five fathoms water. A despatch was at once sent across the country to Bantry Bay to the Great Eastern to come around with all speed, and early the next morning her smoke was seen in the offing. Passing the harbor of Valentia, she proceeded to join the Caroline, which she reached about noon, and at once commenced splicing the massive shore end to her own deep-sea line. This was a work of several hours, so that it was toward evening before all was completed.

Thus, so many had been the delays of the past week, that it had come on to Sunday before the Great Eastern was ready to begin her voyage. This—which some might count a desecration of the holy day—the sailors rather accepted as a good omen. Had the shore end been laid forty-eight hours sooner, the voyage might have begun on Friday, which sailors, who are proverbially superstitious, would have thought an unlucky beginning. But Sunday, in their esteem, is a good day. They like, when a ship is moving out of sight of land, that the last sound from the shore should be the blessed Sabbath bells. If that sacred chime were not heard to-day, at least a Sabbath peace rested on sea and sky. It was a calm summer's evening. The sun was just sinking in the waves, as the Great Eastern, with the two ships of war which waited on either hand, to attend her royal progress, turned their faces to the West, and caught the sudden glory. Says Russell: "As the sun set, a broad stream of golden light was thrown across the smooth billows toward their bows, as if to indicate and illumine the path marked out by the hand of Heaven." What a sacred omen! Had it been the fleet of Columbus sailing westward, every ship's company would have fallen upon their knees on those decks, and burst forth in an Ave Maria to the gentle Mistress of the Seas. But in that manly crew there was many an eye that took in the full beauty of the scene, and many a reverent heart that invoked a benediction.

In other respects the day was well chosen. It was the twenty-third of July. From the beginning, Captain Anderson had wished to sail on the twenty-third of June, or the twenty-second of July, so as to have the full moon on the American coast. He desired also to take advantage of the westerly winds which prevail at that season, for in going against the wind the Great Eastern was steady as a rock. Every expectation was realized. To the big ship the ocean was as an inland lake. The paying-out machinery—the product of so much study and skill—worked beautifully, and as the ship increased her speed, the cable glided into the water with such ease that it seemed but a holiday affair to carry it across to yonder continent. Such were the reflections of all that evening as the long summer twilight lingered on the sea. At midnight they went to sleep, to dream of an easy triumph.

Yet be not too confident. But a few hours had passed before the booming of a gun awoke all on board with the heavy tidings of disaster. The morning breaks early in those high latitudes, and by four o'clock all were on deck, with anxious looks inquiring for the cause of alarm. The ship was lying still, as if her voyage had already come to an end, and electricians, with troubled countenances, were passing in and out of the testing-room, which, as it was always kept darkened, looked like a sick-chamber where some royal patient lay trembling between life and death.

The method used by the electricians to discover a fault is one of such delicacy and beauty as shows the marvellous perfection of the instruments which science employs to learn the secrets of nature. The galvanometer is an invention of Professor Thomson, by which "a ray of light reflected from a tiny mirror suspended to a magnet travels along a scale, and indicates the resistance to the passage of the current through the cable by the deflection of the magnet, which is marked by the course of this speck of light. If the light of the mirror travels beyond the index, or out of bounds, an escape of the current is taking place, and what is technically called a fault has occurred." Such was the discovery on Monday morning. At a quarter past three o'clock the electrician on duty saw the light suddenly glide to the end of the scale and vanish.

Fortunately it was not a fatal injury. It did not prevent signalling through the cable, and a message was at once sent back to the shore, giving notice of the check that had been received. But the electric current did not flow freely. There was a leak at some point of the line which it would not be prudent to pass over. They were now seventy-three miles from shore, having run out eighty-four miles of cable. The tests of the electricians indicated the fault to be ten or a dozen miles from the stern of the ship. The only safe course was to go back and get this on board, and cut out the defective portion. It was a most ungrateful operation thus to be undoing their own work, but there was no help for it.

Such accidents had been anticipated, and before the Great Eastern left England, she had been provided with machinery to be used in case of necessity for picking up the cable. But this proved rather an unwieldy affair. It was at the bow, and as the paying-out machine was at the stern, the ship had to be got round, and the cable, which must first be cut, had to be transferred from one end to the other. This was not an easy matter. The Great Eastern was an eighth of a mile long, and to carry the cable along her sides for this distance, and over her high wheel-houses, was an operation at once tedious and difficult.

But at length the ship's head was brought round, and the end of the cable lifted over the bow, and grasped by the pulling-in machine, and the engine began to puff with the labor of raising the cable from the depths of the ocean. Fortunately they were only in four or five hundred fathoms water, so that the strain was not great. But the engine worked poorly, and the operation was very slow. With the best they could do, it was impossible to raise more than a mile an hour! But patience and courage, though it should take all day and all night!* The Great Eastern did her duty well, steaming slowly back toward Ireland, while the engine pulled, and the cable came up, though reluctantly, from the sea, till on Tuesday morning at seven o'clock, when they had hauled in a little over ten miles, the cause of offence was brought on board. It was found to be a small piece of wire, not longer than a needle, that by some accident (for they did not then suspect a design) had been driven through the outer cover of the cable till it touched the core. There was the source of all the

* " All during the night the process of picking up was carefully carried on, the Big Ship behaving beautifully, and hanging lightly over the cable, as if fearful of breaking the slender cord which swayed up and down in the ocean. Indeed, so delicately did she answer her helm, and coil in the film of thread-like cable over her bows, that she put one in mind of an elephant taking up a straw in its proboscis."—RUSSELL.

mischief. It was this pin's point which pricked the vital cord, opening a minute passage through which the electricity, like a jet of blood from a pierced artery, went streaming into the sea. It was with an almost angry feeling, as if to punish it for its intrusion, that this insignificant and contemptible source of trouble was snatched from its place, the wounded piece of cable was cut off, and a splice made and the work of paying out renewed. But it was four o'clock in the afternoon of Tuesday before they were ready to resume the voyage. A full day and a half had been lost by this miserable piece of wire.

But the vexatious delay was over at last, and the stately ship, once more turning to the West, moved ahead with a steady composure, as if no petty trouble could vex her tranquil mind. Throughout the voyage the behavior of the ship was the admiration of all on board. While her consorts on either side were pitched about at the mercy of the waves, she moved forward with a grave demeanor, as if conscious of her mission, or as if eager to unburden her mighty heart, to throw overboard the great mystery that was coiled up within her, and to cast her burden on the sea.

The electricians, too, were elated, and with reason, at the perfection of the cable as demonstrated by every hour's experience. At intervals of thirty minutes, day and night, tests were passed from ship to shore, and to the delight of all, instead of finding the insulation weakened, it steadily improved as the cable was brought into contact with the cold depths of the Atlantic.

All now went well till Saturday, the twenty-ninth, when a little after noon there was again a cry from the ship, as if once more the cable were wounded and in pain. This time the fault was more serious than before. The electricians looked very grave, for they had struck "dead earth," that is, the insulation was completely destroyed, and the electric current was escaping into the sea.

As the fault had gone overboard, it was necessary to reverse their course, and haul in till the defective part was brought up from the bottom. This time it was more difficult, for they were in water two miles deep. Still the cable yielded slowly to the iron hands that drew it upward; and after working all the afternoon, about ten o'clock at night they got the fault on board. The wounded limb was at once amputated, and joining the parts that were whole, the cable was made new and strong again. Thus ended a day of anxiety. The next morning, which was the second Sabbath at sea, was welcomed with a grateful feeling after the suspense of the last twenty-four hours.

On Monday, the miles of cable that had been hauled up, and which were lying in huge piles upon the deck, were subjected to a rigid examination, to find out where the fault lay. This was soon apparent. Near the end was found a piece of wire thrust through its very heart, as if it had been driven into it. All looked black when this was discovered, for at once it excited suspicions of design. It was remarked that the same gang of workmen were in the tank as at the time of the first fault. Mr. Canning sent for the men, and showing them the cable pierced through with the wire, asked them how it occurred. Every man replied that *it must have been done by design*, even though they accused themselves, as this implied that there was a traitor among them. It seemed hard to believe that any one could be guilty of such devilish malignity. Yet such a thing had been done before in a cable laid in the North Sea, where the insulation was destroyed by a nail driven into it. The man was afterward arrested, and confessed that he had been hired to do it by a rival company. The matter was the subject of a long investigation in the English courts. In the present case there were many motives which might prompt to such an act. The fall in the stock on the London Exchange, caused by a loss of the cable, could hardly be less than half a million sterling. Here was a temptation such as betrays bold, bad men into crime. However, as it was impossible to fix the deed on any one, nothing was proved, and there only remained a painful suspicion of treachery. Against this it was their duty to guard. Therefore it was agreed that the gentlemen on board should take turns in keeping watch in the tank. It was very unpleasant to Mr. Canning thus to set a watch on men, many of whom had been with him in his former cable-laying expeditions, but the best of them admitted the necessity of it, and were as eager as himself to find out the Judas among them.

But accident or villainy, it was defeated this time, and the Great Eastern proudly continued her voyage. Not the slightest check interrupted their progress for the next three days, during which they passed over five hundred miles of ocean. It was now they enjoyed their greatest triumph. They were in the middle of the Atlantic, and thus far the voyage had been a complete success. The ship seemed as if made by Heaven to accomplish this great work of civilization. The paying-out apparatus was a piece of mechanism to excite the enthusiasm of an engineer, so smoothly did its well-oiled wheels run. The strain never exceeded fourteen hundred-weight, even in the greatest depths of the Atlantic. And as for the cable itself, it seemed to come as near perfection as it was possible to attain. As before, the insulation was greatly improved by submergence in the ocean. With every lengthening league it grew better and better. It seems almost beyond belief, yet the fact is fully attested that, when in the middle of the ocean, the communication was so perfect that they could tell at Valentia every time the

Great Eastern rolled.* With such omens of success, who could but feel confident? And when on Monday they passed over a deep valley, where lay "the bones of three Atlantic cables," it was with a proud assurance that they should not add another to the number.

But Wednesday brought a sudden termination of their hopes. They had run out about twelve hundred miles of cable, and were now within six hundred miles of Newfoundland. Two days more would have made them safe, as it would have brought them into the shallow waters of the coast. Thus it was when least expected that disaster came. The record of that fatal day may be given in few words. In the morning, while Mr. Field was keeping watch in the tank, with the same gang of men who had been there when the trouble occurred before, a grating sound was heard, as if a piece of wire had caught in the machinery, and word was passed up to the deck to look out for it; but the caution seems not to have been heard, and it passed over the stern of the ship. Soon after a report came from the testing-room of "another fault." It was not a bad one, since it did not prevent communication with land; and much anxiety might have been saved had a

* So exquisitely sensitive was the copper strand, that as the Great Eastern rolled, and so made the cable pass across the magnetic meridian, the induced current of electricity, incomprehensibly faint as it must have been, produced nevertheless a perceptible deviation of the ray of light on the mirror galvanometer at Foilhommerum.—*London Times.*

THE EXPEDITION OF 1865.

message been sent to Ireland that they were about to cut the cable, in order to haul it on board. But small as the fault was, it could not be left behind. Down on the deep sea-floor was some minute defect, a pin's point in a length measured by thousands of miles. Yet that was enough. Of this marvellous product of human skill, it might in truth be said, that it was like the law of God in demanding absolute perfection. To offend in one point was to be guilty of all.

This new fault, though it was annoying, did not create alarm, for they had been accustomed to such things, and regarded them only as the natural incidents of the voyage. Had the apparatus for pulling in been complete, it could not have delayed them more than a few hours. But this had been the weak point of the arrangements from the beginning—the *bête noire* of the expedition. The only motive power was a little donkey engine, (rightly named,) which puffed and wheezed as if it had the asthma. This was now put in requisition, but soon gave out for want of more steam. While waiting for this a breeze sprang up, which caused the Great Eastern to drift over the cable, by which it was badly chafed, so that when it was hauled in, as the injured part was coming over the bows and was almost within grasp, suddenly it broke and plunged into the sea!

It came without a moment's warning. So unexpected was such a catastrophe, that the gentlemen had gone down to lunch, as it was a little past the hour of noon. But Mr. Canning and Mr. Field stood watching the cable as it was straining upward from the sea, and saw the snapping of that cord, which broke so many hopes. The impression may be better imagined than described. Says a writer on board: "Suddenly Mr. Canning appeared in the saloon, and in a manner which caused every one to start in his seat, said, ' It is all over! It is gone!' then hastened onward to his cabin. Ere the thrill of surprise and pain occasioned by these words had passed away, Mr. Field came from the companion into the saloon, and said, with composure admirable under the circumstances, though his lip quivered and his cheek was blanched, 'The cable has parted and has gone overboard.' All were on deck in a moment, and there, indeed, a glance revealed the truth."

At last it had come—the calamity which all had feared, yet that seemed so far away only a few hours before. Yet there it was—the ragged end on board, torn and bleeding, the other lying far down in its ocean grave.

In America, of course, nothing could be known of the fate of the expedition till its arrival on our shores. But in England its progress was reported from day to day, and as the success up to this point had raised the hopes of all to the highest pitch, the sudden loss of communication with the ship was a heavy blow to public expectation, and gave rise to all sorts of conjectures. At first a favorite theory was, that communication had been interrupted by a magnetic storm. These are among the most mysterious phenomena of nature—so subtle and fleeting as to be almost beyond the reach of science. No visible sign do they give of their presence. No clouds darken the heavens; no thunder peals along the sky. Yet strange influences trouble the air. At this very hour, Professor Airy, the Astronomer Royal at the Observatory at Greenwich, reported a magnetic storm of unusual violence. Said a London paper:

"Just when the signals from the Great Eastern ceased, a magnetic storm of singular violence had set in. Unperceived by us, not to be seen in the heavens, nor felt in the atmosphere, the earth's electricity underwent a mysterious disturbance. The recording instruments scattered about the kingdom, everywhere testified to the fury of this voiceless tempest, and there is every reason to suppose that the confusion of signals at midday on Wednesday was due to the strange and unusual earth-currents of magnetism, sweeping wildly across the cable as it lay in apparently untroubled waters at the bottom of the Atlantic."

Said the Times:

"At Valentia, on Wednesday last, the signals, up to nine A.M., were coming with wonderful distinctness and regularity, but about that time a violent magnetic storm set in. No insulation of a submarine cable is ever so perfect as to withstand the influence of these electrical phenomena, which correspond in some particulars to storms in the ordinary atmosphere, their direction generally being from east to west. Their action is immediately communicated to all conductors of electricity, and a struggle set up between the natural current and that used artificially in sending messages. This magnetic storm affected every telegraphic station in the kingdom. At some the wires were utterly useless; and between Valentia and Killarney the natural current toward the west was so strong along the land lines that it required an addition of five times the ordinary battery power to overcome it. This magnetic storm, which ceased at two A.M. on Friday, was instantly perceptible in the Atlantic cable."

But these explanations, so consoling to anxious friends on land, did not comfort those on board the Great Eastern. They knew, alas! that the cable was at the bottom of the ocean, and the only question was, if any thing could be done to recover it.

Now began a work of which there had been no example in the annals of the sea. The intrepid Canning declared his purpose to grapple for the cable! The proposal seemed wild, dictated by the frenzy of despair. Yet he had fished in deep waters before. He had laid his hand on the bottom of the Mediterranean, but that was a shallow lake compared with the depths into which the Atlantic cable had descended. The ocean is here two and a half miles deep. It was as if an Alpine hunter stood on the summit of Mont Blanc and cast a line into the vale of Chamouni. Yet who shall put bounds to human courage? The expedition was not to be abandoned without a trial of this forlorn hope. There were on board some five miles of wire rope, intended to hold the cable in case it became necessary to cut it and lash it to the buoys, to save it from being lost in a storm. This was brought on deck for another purpose. "And now came forth the grapnels, two five-armed anchors, with flukes sharply curved and tapered to a tooth-like end—the hooks with which the Giant Despair was going to fish from the Great Eastern for a take worth, with all its belongings, more than a million." These huge grappling-irons were firmly shackled to the end of the rope, and brought to the bows and thrown overboard. One splash, and the whole has disappeared in the bosom of the ocean. Down it goes—deeper, deeper, deeper still! For two full hours it continued sinking before it struck the earth, and like a pearl-diver, began searching for its lost treasure on the bottom of the sea. What did it find there? The wrecks of ships that had gone down a hundred years ago, with dead men's bones whitening in the deep sea caves? It sought for something more precious to the interest of civilization than gems and gold.

The ship was now a dozen miles or so from the place of accident. The cable had broken a little after noon, when the sun was shining clear, so that Captains Anderson and Moriarty had just obtained a perfect observation, from which they could tell, within half a mile, the very spot where it had gone down. To reach it now, with any chance of bringing it up, it would be necessary to hook it a few miles from the end. It had been paid out in a line from east to west. To strike it broadside, the ship stood off in the afternoon a few miles to the south. Here the grapnel was thrown over about three o'clock, and struck bottom about five, when the ship began slowly drifting back on her course. All night long those iron fingers were raking the bottom of the deep but grasping nothing, till toward morning the long rope quivered like a fisherman's line when something has seized the end, and the head of the Great Eastern began to sway from her course, as if it felt some unseen attraction. As they began to haul in, the rapidly increasing strain soon rendered it certain that they had got hold of *something*. But what could it be? How did they know it was their lost cable? This question has often been asked. They did not see it. How did they know that it was not the skeleton of a whale, or a mast or spar, the fragment of a wrecked ship? The question is easily answered. If it had been any loose object which was being drawn up from the sea, its weight would have diminished as it came nearer the surface. But on the contrary, the strain, as shown by the dynamometer, steadily *increased*. This could only be from some object lying prone on the bottom. To an engineer the proof was like a mathematical demonstration.

Another fact observed by Captain Anderson was equally decisive:

"The grapnel had caught something at the exact hour when by calculation the ship was known to be crossing the line of the cable; nor had the grapnel upon this or any other occasion even for an instant caught any impediment from the time of its being lowered to the bottom, until the hour indicated by calculation, when the cable ought to be hooked."

Having thus caught the cable, they had good hopes of getting it again, their confidence increasing with every hundred fathoms brought on board. For hours the work went on. They had raised it seven hundred fathoms—or three quarters of a mile—from the bottom, when an iron swivel gave way, and the cable once more fell back into the sea, carrying with it nearly two miles of rope.

The first attempt had failed, but the fact that they had unmistakably caught the prize gave them courage for a second. Preparations were at once begun, but fogs came on and delayed the attempt till Monday, when it was repeated. The grapnel caught again. It was late in the afternoon when it got its hold, and the work of pulling in was kept up all night. But as the sea was calm and the moon shining brightly, all joined in it with spirit, feeling elated with the hope of triumph on the morrow.

That was not to be; but each attempt seemed to come nearer and nearer to victory. This time the cable was drawn up a full mile from the bottom, and hung suspended a mile and a half below the ship. Had the rope been strong enough, it might have been brought on board. But again a swivel gave way, and the cable, whose sleep had been a second time disturbed, sought its ocean bed.

These experiments were fast using up the wire rope, and every expedient had to be resorted to, to piece it out and to give it strength. Each shackle and swivel was replaced by new bolts, and the capstan was increased four feet in diameter, by being belted with enormous plates of iron, to wind the rope around it, if the picking-up machinery should fail. This gave full work to all the mechanics on board. The ship was turned into a very cave of Vulcan, presenting at night a scene which might well take the eye of an artist, and which Russell thus describes:

"The forge fires glared on her decks, and there, out in the midst of the Atlantic, anvils rang and sparks flew; and the spectator thought of some village far away, where the blacksmith worked, unvexed by cable anxieties and greed of speedy news. As the blaze shot up, ruddy, mellow and strong, and flung arms of light aloft and along the glistening decks, and then died into a red centre, masts, spars, and ropes were for the instant touched with a golden gleaming, and strange figures and faces were called out from the darkness—vanished, glinted out again—rushed suddenly into foreground of bright pictures, which faded soon away— flickered—went out—as they were called to life by its warm

THE EXPEDITION OF 1865.

breath, or were buried in the outer darkness! Outside all was obscurity, but now and then vast shadows, which moved across the arc of the lighted fog-bank, were projected far away by the flare; and one might well pardon the passing mariner, whose bark drifted him in the night across the track of the great ship, if, crossing himself, and praying with shuddering lips, he fancied he beheld a phantom ship freighted with an evil crew, and ever after told how he had seen the workshops of the Inferno floating on the bosom of the ocean."

While preparing for a third attempt, the ship had been drifting about, sometimes to a distance of thirty or forty miles, but it had marked the course where the cable lay by two buoys thrown over about ten miles apart, each bearing a flag which might be seen at a distance, and so easily came back to the spot. On Thursday morning all was ready, and the line was cast as before, but after some hours of drifting, it was evident that the ship had passed over the cable without grappling. The line was hauled in, and the reason at once appeared. One of the flukes had caught in the chain, so that it could not strike its teeth into the bottom. This was cleared away, and the rope prepared for a fourth and final attempt.

It was at noon of Friday that the grapnel went overboard for the last time. By four o'clock it had caught, and the work of hauling in recommenced. Again the cable was brought up nearly eight hundred fathoms, when the rope broke, carrying down two miles of its own length, and with it the hopes of the Atlantic Telegraph for the present year.

Their resources were exhausted. For nine days and nights, for the work never stopped for light or darkness, had the great ship kept moving round and round like some mighty bird of the sea, with her eye fixed on the place where her treasure had gone down, and striving to wrest it from the hand of the spoiler. Three times had they grasped the prize, and each time failed to recover it, only for want of ropes strong enough to bring it on board. *The cable itself never broke.* This proof of its strength was a good omen for future success.

But for the present all was over. The attempt must be abandoned for the year 1865, but not for ever; and with this purpose in her constant mind, the Great Eastern swung sullenly around, and turned her imperial head toward England, like a warrior retiring from the field—not victorious, nor yet defeated and despairing, but with her battle-flag still flying, and resolved once more to attempt the conquest of the sea.

CHAPTER XV.

PREPARING TO RENEW THE BATTLE.

THE expedition of 1865, though not an immediate success, had the moral effect of a victory, as it confirmed the most sanguine expectations of all who embarked in it. The great experiment made during those four weeks at sea, had demonstrated many points which were most important elements in the problem of Ocean Telegraphy. These are summed up in the following paper, which was signed by persons officially engaged on board the Great Eastern:

1. It was proved by the expedition of 1858, that a Submarine Telegraph Cable could be laid between Ireland and Newfoundland, and messages transmitted through the same.

By the expedition of 1865 it has been fully demonstrated:

2. That the insulation of a cable improves very much after its submersion in the cold deep water of the Atlantic, and that its conducting power is considerably increased thereby.

3. That the steamship Great Eastern, from her size and constant steadiness, and from the control over her afforded by the joint use of paddles and screw, renders it safe to lay an Atlantic Cable in any weather.

4. That in a depth of over two miles four attempts were made to grapple the cable. In three of them the cable was caught by the grapnel, and in the other the grapnel was fouled by the chain attached to it.

5. That the paying-out machinery used on board the Great Eastern worked perfectly, and can be confidently relied on for laying cables across the Atlantic.

6. That with the improved telegraphic instruments for long submarine lines, a speed of more than eight words per minute can be obtained through such a cable as the present Atlantic between Ireland and Newfoundland, as the amount of slack actually paid out did not exceed fourteen per cent, which would have made the total cable laid between Valentia and Heart's Content nineteen hundred miles.

7. That the present Atlantic Cable, though capable of bearing a strain of seven tons, did not experience more than fourteen hundred-weight in being paid out into the deepest water of the Atlantic between Ireland and Newfoundland.

8. That there is no difficulty in mooring buoys in the deep water of the Atlantic between Ireland and Newfoundland, and that two buoys even when moored by a piece of the Atlantic Cable itself, which had been previously lifted from the bottom, have ridden out a gale.

9. That more than four nautical miles of the Atlantic Cable have been recovered from a depth of over two miles, and that the insulation of the gutta-percha-covered wire was in no way whatever impaired by the depth of water or the strains to which it had been subjected by lifting and passing through the hauling-in apparatus.

PREPARING TO RENEW THE BATTLE.

10. That the cable of 1865, owing to the improvements introduced into the manufacture of the gutta-percha core, was more than one hundred times better insulated than cables made in 1858, then considered perfect and still working.

11. That the electrical testing can be conducted with such unerring accuracy as to enable the electricians to discover the existence of a fault immediately after its production or development, and very quickly to ascertain its position in the cable.

12. That with a steam-engine attached to the paying-out machinery, should a fault be discovered on board whilst laying the cable, it is possible that it might be recovered before it had reached the bottom of the Atlantic, and repaired at once.

    S. CANNING, Engineer-in-Chief, Telegraph Construction
      and Maintenance Company.
    JAMES ANDERSON, Commander of the Great Eastern.
    HENRY A. MORIARTY, Staff Commander, R. N.
    DANIEL GOOCH, M.P., Chairman of "Great Ship Co."
    HENRY CLIFFORD, Engineer.
    WILLIAM THOMSON, LL.D., F.R.S., Prof. of Natural Philosophy in the University of Glasgow.
    CROMWELL F. VARLEY, Consulting Electrician Electric and International Telegraph Co.
    WILLOUGHBY SMITH.
    JULES DESPECHER.

This was a grand result to be attained in one short month; and if not quite so gratifying as to have the cable laid at once, and the wire in full operation, yet as it settled the chief elements of success, the moral effect was next to that of an immediate triumph. All who were on that voyage felt a confidence such as they had never felt before. They came back, not desponding and discouraged, but buoyant with hope, and ready at once to renew the attempt.

This confidence appeared at the first meeting of directors. The feeling was very different from that after the return of the first expedition of 1858. So animated were they with hope, and so sure of success the next time, that all felt that one cable was not enough, they must have two, and so it was decided to take measures not only to raise the broken end of the cable and to complete it to Newfoundland, but also to construct and lay an entirely new one, so as to have a double line in operation the following summer.

The contractors, partaking the general confidence, came forward promptly with a new offer even more liberal than that made before. They proposed to construct a new line, and to lay it across the Atlantic for half a million sterling, which was estimated to be the actual cost to them, reserving all compensation to themselves to depend on success. If successful, they were to receive twenty per cent. on the cost, or one hundred thousand pounds, to be paid in shares of the Company. They would engage, also, to go to sea fully prepared to raise the broken end of the cable now in mid-ocean, and with a sufficient length, including that on board the Great Eastern, to complete the line to Newfoundland. Thus the company would have two cables instead of one.

In this offer the contractors assumed a very large risk. They now went a step further, and in the contingency of the capital not being raised otherwise, they offered *to take it all themselves*—to lay the line at their own risk, and to be paid only in the stock of the Company, which, of course, must depend for its value on the success of the next expedition. It was finally resolved to raise six hundred thousand pounds of new capital by the issue of a hundred and twenty thousand shares of five pounds each, which should be preferential shares, entitled to a dividend of twelve per cent. before the eight per cent. dividend to be paid on the former preference shares, and the four per cent. on the ordinary stock. This was offering a substantial inducement to the public to take part in the enterprise, and it was thought with reason that this fresh issue of stock, though it increased the capital of the Company, yet as it was all to be employed in forwarding the great work, would not only create new property, but give value to the old. The proposal of the manufacturers was therefore at once accepted by the Directors, and the work was instantly begun. Thus hopeful was the state of affairs when Mr. Field returned to America in September.

But he was never easy to be long out of sight of his beloved cable, and so three months after he went back to England, reaching London on the twenty-fourth of December. He came at just the right moment, for the Atlantic Telegraph was once more in extremity. Only two days before the Attorney-General of England had given a written opinion that the Company *had no legal right* to issue new twelve per cent. preference shares, and that such issue could only be authorized by an express act of Parliament. This was a fatal decree to the Company. It was the more unexpected, as, before offering the twelve per cent. capital, they had been fortified by the opinions of several eminent lawyers and solicitors in favor of the legality of their proceedings. It invalidated not only what they were going to do, but what they had done already. Hence, as the effect of this decision, all the works were stopped, and the money which had been paid in was returned to the subscribers.

This was a new dilemma, out of which it was not easy to find a way of relief. Parliament was not in session, Lords and Commons being away in the country keeping the Christmas holidays. Even if it had been, the time for applying to it had passed, as a notice of any private bill to be introduced must be given before the thirtieth of November, which was gone a month ago. To wait for an act of Parliament, therefore, would inevitably postpone the laying of the cable for another year. So disheartening was the prospect at the close of 1865.

PREPARING TO RENEW THE BATTLE.

But they had seen dark days before, and were not to give it up without a new effort. Happily, the cause had strong friends to stand by it even in this crisis of suspended animation.

One of these to whom Mr. Field now went for counsel, was Mr. (afterward Sir) Daniel Gooch, M.P., a gentleman well known in London, as one of the class of engineers formed in the school of Stephenson and Brunel, who had risen to the position of great capitalists, and who, by their enterprise and wealth, had done so much to develope the resources of England. He was Chairman of the Great Western Railway, and had more faith in enterprises on the land than on the sea. It was a long time before he could believe in the possibility of an Atlantic Telegraph. Though a man of large fortune, and a personal friend of Mr. Field, the latter had never prevailed on him to subscribe a single pound. But he went out on the expedition of '65, as chairman of the company that owned the Great Eastern; and what he then saw convinced him. He came back fully satisfied; he knew it could be done, and was ready to prove his faith by his works. Consulting on the present difficulty, he suggested that the only relief was to organize a new Company, which should assume the work, and which could issue its own shares and raise its own capital. This opinion was confirmed by the eminent legal authority of Mr. John Horatio Lloyd. To such a Company Mr. Gooch said he would subscribe £20,000; Mr. Field put down £10,000.

Next, he betook himself to that prince of English capitalists, Mr. Thomas Brassey, who heard from his lips for the first time, that the affairs of the Atlantic Telegraph Company had suddenly come to a standstill. At this he was much surprised, but instantly cheered his informer by saying: "Don't be discouraged; go down to the Company, and tell them to go ahead, and whatever the cost, I will bear one tenth of the whole." Who *could* be discouraged with such a Richard the Lion-hearted to cheer him on?

Meetings were called of the Directors of both the Atlantic Company and the Telegraph Construction and Maintenance Company; and frequent conferences were held between them. The result was the formation of a new company called the Anglo-American Telegraph Company, with a capital of £600,000, which contracted with the Atlantic Company to manufacture and lay down a cable in the summer of 1866, for doing which it was to be entitled to what virtually amounted to a preference dividend of twenty-five per cent: as a first claim was secured to them by the latter company upon the revenue of the cable or cables (after the working expenses had been provided for) to the extent of £125,000 per annum; and the New-York, Newfoundland, and London Telegraph Company undertook to contribute from its revenue a further annual sum of £25,000, on condition that a cable should be at work during 1866; an agreement to this effect having been signed by Mr. Field, subject to ratification by the Company in New York, which was obtained as soon as the steamer could cross the ocean and bring back the reply.

The terms being settled, it remained only to raise the capital. The Telegraph Construction and Maintenance Company led off with a subscription of £100,000. This was followed by the names of ten gentlemen, who put down £10,000 apiece. Of these Mr. Gooch declared his willingness to increase his subscription of £10,000 to £20,000, while Mr. Brassey would put down £60,000, if it were needed. Mr. Henry Bewley, of Dublin also, who was already a large owner of the Atlantic stock, declared his readiness to add £20,000 more. But this was not necessary: and so they all stood at £10,000. The names of these ten subscribers deserve to be given, as showing who stood forward to save the cause in this crisis of its fate. They were: Henry Ford Barclay, Henry Bewley, Thomas Brassey, A. H. Campbell, M.P., George Elliot, Cyrus W. Field, Richard Atwood Glass, Daniel Gooch, M.P., John Pender, M.P., and John Smith. There were four subscriptions of £5,000: by Thomas Bolton and Sons, James Horsfall, A Friend of Mr. Daniel Gooch, M.P., and John and Edwin Wright; one of £2,500 by John Wilkes and Sons; three of £2,000 by C. M.

Lampson, J. Morison, and Ebenezer Pike; and two of £1,000 by Edward Cropper, and Joseph Robinson,— making in all £230,500.

These were all private subscriptions made before even the prospectus was issued, or the books opened to the public. After such a manifestation of confidence, the whole capital was secured within fourteen days. This was a great triumph, especially at a time of general depression in commercial affairs in England.

And now once more the work began. No time was to be lost. It was already the first of March, and but four months remained to manufacture sixteen hundred and sixty nautical miles of cable, and to prepare for sea. But the obstacles once cleared away, all sprang to their work with new hope and vigor.

In the cable to be made for the new line, there was but little change from that of the last year, which had proved nearly perfect. Experience, however, was constantly suggesting some improvement; and while the general form and size were retained, a slight change in the outer covering was found to make the cable both lighter and stronger. The iron wires were *galvanized*, which secured them perfectly from rust or corrosion by salt water. Thus protected, they could dispense with the preservative mixture of the former year. This left the cable much cleaner and whiter. Instead of its black coat, it had the fresh, bright appearance of new rope. It had another advantage.

PREPARING TO RENEW THE BATTLE.

As the tarry coating was sticky, slight fragments of wire might adhere to it, and do injury, a danger to which the new cable was not exposed. At the same time, galvanizing the wires gave them greater ductility, so that in the case of a heavy strain the cable would stretch longer without breaking. By this alteration it was rendered more than four hundred-weight lighter per mile, and would bear a strain of nearly half a ton more than the one laid the year before.

The machinery also was perfected in every part, to withstand the great strain which might be brought upon it in grappling and lifting the cable from the great depths of the Atlantic. This necessitated almost a reconstruction of the machinery, together with engines of greater power, applied both to the gear for hauling in forward and that for paying out aft. Thus, in case of a fault, the motion of the ship could be easily reversed, and the cable hauled back by the paying-out machinery, without waiting for the long and tedious process of bringing the cable round from the stern to the bow of the ship.

But the most marvellous improvement had been in the method of testing the cable for the discovery of faults. In the last expedition, a grave omission had been in the long intervals during which the cable was left without a test of its insulation. Thus, from thirty to thirty-five minutes in each hour it was occupied with tests of minor importance, which would not indicate the existence of a fault, so that, if a fault occurred on ship-board, it might pass over the stern, and be miles away before it was discovered. But now a new and ingenious method was devised by Mr. Willoughby Smith, by which the cable will be tested *every instant*. The current will not cease to flow any more than the blood ceases to flow in human veins. The cord is vital in every part, and if touched at any point it reveals the wound as instinctively as the nerves of a living man flash to the brain a wound in any part of the human frame.

The process of detecting faults is too scientific to be detailed in these pages. We can only stand in silent wonder at the result, when we hear it stated by Mr. Varley, that the system of testing is brought to such a degree of perfection, that skilful electricians can point out minute faults with an unerring accuracy "even when they are so small that they would not weaken the signals through the Atlantic cable one millionth part!"

Another marvellous result of science was the exact report obtained of the state of that portion of the cable now lying in the sea. The electricians at Valentia were daily experimenting on the line which lay stretched twelve hundred and thirteen miles on the bottom of the deep, and pronounced it intact. Not a fault could be found from one end to the other. As when a master of the organ runs his hands over the keys, and tells in an instant if it be in perfect tune, so did these skilful manipulators, fingering at the end of this mightier instrument, declare it to be in perfect tone, ready to whisper its harmonies through the seas. At the same time, the ten hundred and seventy miles of cable left on board the Great Eastern were pronounced as faultless as the day they had been shipped on board.

With such conclusions of science to animate and inspire them, the great task of manufacturing nearly seventeen hundred miles of cable once more began. And while this work went on, the Great Eastern, that had done her part so well before, again opened her sides, and the mysterious cord was drawn into her vast, dark, silent womb, from which it was to issue only into the darker and more silent bosom of the deep.

CHAPTER XVI.

VICTORY AT LAST.

IN these pages we have led our readers through twelve long years, and have had to tell many a tale of disaster and defeat. It is now our privilege to tell of triumphant success. Victory has come at last, but not by the chance of fortune, but by the utmost efforts of man, by the union of science and skill with indomitable perseverance. The failure of the last year was a sad disappointment; but so far from damping the courage of those embarked in the enterprise, it only roused them to a more gigantic effort. They were now to prepare for a fifth expedition. In this they set themselves to anticipate every possible emergency, and to combine the elements of success so as to render failure impossible.

The Great Eastern herself, which they had come to regard with a kind of fondness, a feeling of affection and pride, as the ark that was to bear their fortunes across the deep, was made ready for her crowning achievement. For months Captain Anderson and Mr. Halpin, his chief officer, worked day and night to get her into perfect trim. She had become sadly fouled in her many voyages. As she swam the seas, a thousand things clung to her as to a floating island, till her hull was encrusted with mussels and barnacles two feet thick, and long seaweed flaunted from her sides. Like a brave old war-horse, long neglected, she needed a thorough grooming, to have her hair combed and her limbs well rubbed down, to fit her to take the field. But it was not an easy matter to get under the huge creature, to give her such a dressing. Yet Captain Anderson was equal to the emergency. He contrived a simple instrument by which every part of her bottom was raked and scrubbed. Getting rid of this rough, shapeless mass would make her feel easy and comfortable at sea, and add at least a knot an hour to her speed.

The boilers too were thoroughly cleansed and repaired in every part, and the paddle-engines were so arranged that in five minutes they could be disconnected, so that by going ahead with one and backing with the other, the ship could be held perfectly at rest or be turned around in her own length, a very important matter when they should come to fish in deep waters for the broken end of the cable. To prepare for this, she was armed with chains and ropes and irons of the most formidable kind. For grappling the cable, she took on board twenty miles of rope, which would bear a strain of thirty tons, probably the largest fishing-line used since the days of Noah!

The cable was manufactured at the rate of twenty miles a day, and as fast as delivered and found perfect, was coiled on board. And now the electricians tried their skill to outdo all that they had done before. As Captain Anderson observed, it seemed as if never had so much brain power been concentrated on the problem of success. The cable itself furnished the grandest subject of experiment. As every week added more than a hundred miles to its length, there was constant opportunity to try the electric current on longer distances and with new conditions. The results obtained showed the rapid and marvellous progress of electrical science. Said The Times:

"The science of making, testing, and laying cables has so much improved that an undetected fault in an insulated wire has now become literally impossible, while so much are the instruments for signalling improved, that not only can a slight fault be disregarded if necessary, but it is even easy *to work through a submarine wire with a foot of its copper conductor stripped and bare to the water.* This latter result, astonishing as it may appear, has actually been achieved for some days past with the whole Atlantic cable on board the Great Eastern. Out of a length of more than one thousand seven hundred miles, a coil has been taken from the centre, the copper conductor stripped clean of its insulation for a foot in length, and in this condition lowered over the vessel's side till it rested on the ground. Yet through this the clearest signals have been sent—so clear, indeed, as at one time to raise the question whether it would not be worth while to grapple for the first old Atlantic cable ever laid, and with these new instruments working gently through it for a year or so, at least make it pay cost."

As other things were on the same gigantic scale, by the time the big ship had her cargo and stores on board, she was well laden. Of the cable alone there were two thousand four hundred miles, coiled in three immense tanks as the year before. Of this seven hundred and forty-eight miles were a part of the cable of the last expedition. The tanks alone, with the water in them, weighed over a thousand tons; and the cable which they held, four thousand tons more; besides which she had to carry eight thousand five hundred tons of coal and five hundred tons of telegraph stores, making fourteen thousand tons, besides engines, rigging, etc., which made nearly as much more. So enormous was the burden, that it was thought prudent not to take on board all her coal before she left the Medway, especially as the channel was winding and shallow. It was therefore arranged that about a third of her coal should be taken in at Berehaven, on the south-west coast of Ireland. With this exception, her lading was complete.

The time for departure had been fixed for the last day of June, and so admirable had been the arrangements, and such the diligence of all concerned, that exactly at the hour of noon, she loosened from her moorings, and began to move. It was well that she had not on board her whole cargo; for as it was, she drew nearly thirty-two feet. Never had any keel pressed so deep in those waters. It required skilful handling to get her safely to the sea. Gently and softly she floated down, over bars where she almost grazed the sand, where but a few inches lifted her enormous hull above the river's bed. But at length the rising tide bears her safely over, and she is afloat in the deeper waters of the Channel. At first the sea did not give her a very gracious welcome. The wind was dead ahead, and the waves dashed furiously against her; but she kept steadily on, tossing their spray on high, as if they had struck against the rocks of Eddystone lighthouse. In four or five days she had passed down the Irish coast, and was quietly anchored in the harbor at Berehaven, where she was soon joined by the other vessels of the squadron.

The Telegraph fleet was not the same as that of the last year. The Government could spare but a single ship; but the Terrible, which had accompanied the Great Eastern on the former expedition, was still there to represent the majesty of England. The William Corry, a vessel of two thousand tons, bore the ponderous shore end, which was to be laid out thirty miles from the Irish coast, while the Albany and the Medway were ships chartered by the Company. The latter carried several hundred miles of the last year's cable, besides one of heavier proportions, ninety miles

VICTORY AT LAST. 311

long, to be stretched across the mouth of the Gulf of St. Lawrence.

While the Great Eastern remained at Berehaven, to take in her final stores of coal, the William Corry proceeded around the coast to Valentia to lay the shore end. She arrived off the harbor on the morning of

SHORE END—EXACT SIZE.

Saturday, the seventh of July, and immediately began to prepare for her heavy task. This shore end was of tremendous size, weighing twenty tons to the mile. It was by far the strongest wire cable ever made, and in short lengths was stiff as an iron bar. As the year before, the cable was to be brought off on a bridge of boats reaching from the ship to the foot of the cliff.

All the fishermen's boats were gathered from along the shore, while H. M. S. Raccoon, which was guarding that part of the coast, sent up her boats to help, so that, as they all mustered in line, there were forty of them, making a long pontoon-bridge; and Irish boatmen with eager looks and strong hands were standing along the line, to grasp the ponderous chain. All went well, and by one o'clock the cable was landed, and its end brought up the cliff to the station. The signals were found to be perfect, and the William Corry then slowly drew off to sea, unlimbering her stiff shore end, till she had cast over the whole thirty miles. At three o'clock next morning she telegraphed through the cable that her work was done, and she had buoyed the end in water a hundred fathoms deep. Describing the scene, the correspondent of the London News says:

"In its leading features it presented a striking difference to the ceremony of last year. Earnest gravity and a deep-seated determination to repress all show of the enthusiasm of which everybody was full, was very manifest. The excitement was below, instead of above, the surface. Speech-making, hurrahing, public congratulations, and vaunts of confidence were, as it seemed, avoided as if on purpose. There was something far more touching in the quiet and reverent solemnity of the spectators yesterday than in the slightly boisterous joviality of the peasantry last year. Nothing could prevent the scene being intensely dramatic, but the prevailing tone of the drama was serious instead of comic and triumphant. The old crones in tattered garments

VICTORY AT LAST. 313

who cowered together, dudheen in mouth, their gaudy colored shawls tightly drawn over head and under the chin—the barefooted boys and girls, who by long practice walked over sharp and jagged rocks, which cut up boots and shoes, with perfect impunity—the men at work uncovering the trench, and winding in single file up and down the hazardous path cut by the cablemen in the otherwise inaccessible rock—the patches of bright color furnished by the red petticoats and cloaks—the ragged garments, only kept from falling to pieces by bits of string and tape—the good old parish priest, who exercises mild and gentle spiritual sway over the loving subjects of whom the ever-popular Knight of Kerry is the temporal head, looking on benignly from his car—the bright eyes, supple figures, and innocent faces of the peasant lasses, and the earnestly hopeful expression of all—made up a picture impossible to describe with justice. Add to this, the startling abruptness with which the tremendous cliffs stand flush out of the water, the alternations of bright wild flowers and patches of verdure with the most desolate barrenness, the mountain sheep indifferently cropping the short, sweet grass, and the undercurrent of consciousness of the mighty interests at stake, and few scenes will seem more important and interesting than that of yesterday."

As the ships are now ready for sea, and all who are to embark have come on board, we may look about us at the personnel of the expedition. Who are here? We recognize many old familiar faces, that we have seen in former campaigns—gallant men who have had many a sea-fight in this peaceful war. First, the eye seeks the tall form of Captain Anderson. There he is, modest and grave, of few words, but seeing every thing, watching every thing, and ruling every thing with a quiet power. And there is his second officer, Mr. Halpin, who keeps a sharp lookout after the crew, to see that every man does his duty. While he thus keeps watch of all on board, Staff Commander Moriarty, R. N., comes on deck, with instruments in hand, to look after the heavenly bodies, and reckon the ship's latitude and longitude. This is an old veteran in the service, who has been in all the expeditions, and it would be quite "improper," even if it were possible, for a cable to be laid across the Atlantic without his presence and aid. And here comes Mr. Canning, the engineer, whose deep-sea soundings, the last year, were on a scale of such magnitude, and who, if he cannot well dive deeper, means to pull stronger the next time. That slight form yonder is Professor Thomson, of Glasgow, a man who in his knowledge of the subtle element to be brought into play, and the enthusiasm he brings to its study, is the very genius of electrical science; and this is Mr. Varley, who seems to have the lightning in his fingers, and to whom the world owes some marvellous discoveries of the laws of electricity. Mr. Willoughby Smith, a worthy associate in these studies and discoveries, goes out on the ship as electrician.

And here is Mr. Glass, the managing director of the Telegraph Construction and Maintenance Company, which has undertaken by contract to manufacture this cable and lay it safely across the ocean; and Mr. Gooch, chairman of the company that owns the Great Eastern—two gentlemen to whom the Atlantic Telegraph is under the greatest obligation, since it was they who, six months before, when the project seemed in danger of being given up or postponed for years, took Mr. Field by the hand, and cheered him on to a last effort. Blessings on their hearts of oak! Mr. Gooch accompanies the ship, while Mr. Glass, keeping Mr. Varley at his side as electrician, remains on shore, to receive reports of the daily progress of the expedition, and to issue his orders. What a post of observation was that telegraph house on the cliffs of Valentia! It commanded a far broader horizon than the top of Fiesolé, from which Galileo looked down on the valley of the Arno, and up at the stars. Was there ever a naval commander favored with a power of vision that could sweep the boundless sea? What would Nelson have said, if he had had a spy-glass with which he could watch ships in action two thousand miles away, and issue his orders to a fleet on the other side of the ocean? With such a long range, he might almost have fought the Battle of the Nile from his home in England.

Standing on such a spot, and surrounded by such men, representing the capital, the science, and the skill of England, with all those gallant ships in sight, one's heart might well beat high. But there were other reflections that saddened the hour, and caused some at least to look once more on the rocks of Valentia with deep emotion. Some of their old companions-in-arms had fallen out of the ranks, while the battle was not yet won. Brett, Mr. Field's first friend in England, was in his grave. Beyond the Atlantic, Captains Hudson and Berryman slept the sleep that knows no waking. They were not forgotten by their survivors, who mourned that those who had toiled with them in former days, were not here to share their triumph.

The feeling, therefore, of many on this occasion, was not one elate with pride and hope, but subdued by serious thoughts and tender memories. In harmony with this feeling, and with the great work which they were about to undertake, it was proposed that before the expedition sailed they should hold a solemn religious service.

Was there ever a fitter place or a fitter hour for prayer than here, in the presence of the .great sea to which they were about to commit their lives and their precious trust? The first expedition ever sent forth had been consecrated by prayer. On that very spot, nine years before, all heads were uncovered and all forms bent low, at the solemn words of supplication; and there had the Earl of Carlisle—since gone to his honored grave—cheered them on with high religious hopes, describing the ships which were sent forth on such a mission, as "beautiful upon the waters as were the feet upon the mountains of them that publish the gospel of peace."

In such a spirit two of the directors—Mr. Bevan, of London, and Mr. Bewley, of Dublin—sent invitations to a number of persons to meet at Valentia, as the expedition was about to sail, and commend it to the favor of Almighty God. Captain Anderson had greatly desired to be with them at this parting service, but the ships were at Berehaven, and they were just embarking for sea. But though the officers could not be present, a large company came together. Said an Irish paper : "Men of different religious denomination, and of various professions in life—Irishmen, Englishmen, and Scotchmen—joined in such a service as has never been held in this island." It was a scene long to be remembered, as they bowed together before the God and Father of all. Their brethren, who were about to go down to the sea in ships, felt their dependence on a Higher Power. Their preparations were complete. All that man could do was done. They had exhausted every resource of science and skill. The issue now remained with Him who controls the winds and waves. Therefore was it most fit that, at the very moment of embarking, those who remained behind should, as it were, kneel upon the cliff, and, with outstretched hands, commit them to Him who alone spreadeth out the heavens and ruleth the raging of the sea.

In all this there is something of antique stamp, something which makes us think of the sublime men of an earlier and better time; of the Pilgrim Fathers kneeling on the deck of their little ship at Leyden, as they were about to seek a refuge and a home in the forests of the New World; and of Columbus and his companions celebrating a solemn service before their departure from Spain. And so with labor and with prayer did this great expedition go forth once more from the shores of Ireland, bearing the hopes of science and of civilization—with courage and skill looking out from the bow across the stormy waters, and a religious faith, like that of Columbus, standing at the helm.

On Friday morning, the thirteenth of July, the fleet finally bade adieu to the land. Was Friday an unlucky day? Some of the sailors thought so, and would have been glad to leave a day before or after. But Columbus sailed on Friday, and discovered the New World on Friday; and so this expedition put to sea on Friday, and, as a good Providence would have it, reached land on the other side of the Atlantic on the same day of the week! As the ships disappeared below the horizon, Mr. Glass and Mr. Varley went up on their watch-tower—not to look, but to listen for the first voice from the sea. The ships bore away for the buoy where lay the end of the shore line; but the weather was thick and foggy, with frequent bursts of rain, and they could not see far on the water. For an hour or two they went sailing round and round, like sea-gulls in search of prey. At length the Albany caught sight of the buoy tossing on the waves, and, firing a signal gun, bore down straight upon it. The cable was soon hauled up from its bed, a hundred fathoms deep, and brought over the stern of the Great Eastern; and the watchers on shore, who had been waiting with some impatience, saw the first flash, and Varley read, "Got the shore end—all right—going to make the splice." Then all was still, and they knew that that delicate operation was going on. Quick, nimble hands tore off the covering from some yards of the shore end of the main cable, till they came to the core; then, swiftly unwinding the copper wires, they laid them together, twining them as closely and carefully as a silken braid. Thus stripped and bare this new-born child of the sea was wrapped in swaddling-clothes, covered up with many coatings of gutta-percha, and hempen rope, and strong iron wires, the whole bound round and round with heavy bands, and the splicing was complete. Signals were now sent through the whole cable on board the Great Eastern and back to the telegraph-house at Valentia, and the whole length, two thousand four hundred and forty nautical miles, was reported perfect. And so with light hearts they bore away. It was a little after three o'clock. As they turned to the west, the following was the "order of battle": the Terrible went ahead, standing off on the starboard bow, to keep other vessels out of the course; the Medway was on the port, and the Albany on the starboard quarter, ready to pick up or let go a buoy, or to do other work that might be required. All these ships were to keep their allotted positions, within signalling distance of the Great Eastern, and at any time that she was heard firing guns, they were to close in with her to render assistance. Their course lay thirty miles to the south of that of the last year, so that there could be no danger, in fishing for the old cable, of disturbing the new.

Dr. Russell, the brilliant historian of the Expedition of 1865, was not on board the Great Eastern this year. He had left England a few weeks before for the scene of the war in Germany. His place was supplied by Mr. John C. Deane, the Secretary of the Anglo-American Company, whose "Diary of the Expedition" furnishes a faithful record of the incidents of this memorable voyage. If the story be not quite so thrilling as that of the year before, it is because it has not to tell of such fatal accidents. It has the monotony of success. A few pages from this diary, giving its most important portions, will render this narrative complete.

The voyage began with good weather and every omen of success. Friday, indeed, was a day of fog

VICTORY AT LAST.

and rain. At the very time they were making the splice with the shore end, the rain was pouring on the deck. But in a few hours it cleared off, and Saturday and Sunday, Mr. Field writes in his journal, " Weather fine ; " and Monday, " Calm, beautiful day. Signals perfect." Owing to the improved system adopted by the chief electrician, communication with the shore was kept up even while the tests for insulation were going on.*

* The new method is thus explained by Mr. Deane :

" The fundamental difference between last year's system of testing and that of the present expedition is, that now all the ordinary tests for continuity may be made simultaneously with the test for insulation, which is not interrupted at all ; whereas, last year, during half the time spent in laying the cable, the insulation test was wholly neglected.

" Last year, each hour was divided into four parts. The first half of the hour was spent in testing for insulation. During the second half, which was divided into three periods of ten minutes each, tests were made to ascertain the resistance of the conductor and to prove the continuity of the same. All these tests were of such a nature as to afford no criterion whatever of the state of the insulation during their continuance, so that during the half of each hour, or, in other words, during half the time spent in laying the cable, the insulation test was neglected. Also, while the insulation test was being made, there was no means of communicating with the shore, as the observations were taken on board only. This year, a test for insulation is constantly kept on, and, by Mr. Willoughby Smith's arrangement, corresponding observations are made both on ship and shore. At stated times during the hour, the continuity test is made at the shore station by means of a condenser applied to the conductor of the cable. The effect of this is to increase the deflection on the ship's insulation galvanometer, thus serving as a continuity test. Communications from shore to ship are also made by these means. The ship can send signals to the shore by

Every possible precaution was taken to guard against such accidents as had marred the success of the year before. Remembering how small a thing had sufficed to puncture the cable, the men in the tank were not allowed to wear boots or shoes with nails in their heels, but were cased from head to foot in canvas dresses, drawn over their ordinary sailor costume, and, with slippers on their feet, they glided about softly as ghosts. But we turn to Mr. Deane's diary for a record of the progress from day to day:

"Sunday, July 15.—All through yesterday the paying-out machinery worked so smoothly—the electrical tests were so perfect—the weather was so fine, that fresh confidence in the ultimate result has been naturally inspired. The recollection, however, of the reverses of the expedition of 1865 is always before those who have the greatest reliance on success; and there is a quiet repose about the manner of the chief practical men on board, which is an earnest that they will not allow themselves to be carried away by the smoothness of twenty-four hours' events. The convoy kept their position accurately during the day. The Terrible signalled that a man had fallen overboard. Her cutter was speedily lowered. The sailor had, however, laid hold of a rope thrown to him from the frigate, before the boat reached him.

simply reversing the current for certain lengths of time, answering to some understood code, or by increasing and diminishing the tension of the line, according to a preärranged plan. All these operations may be performed without interrupting the insulation test, except for a few seconds while the current is being reversed. So far for the new system in the electrical room as compared with last year."

"Monday.—Still everything going on well. The sea like a mill-pond. The paying out of the cable from the after tank progressing with uniformity and steadiness, and the electrical tests perfect.

"Our track is about thirty miles to the south of that of last year, and at that distance we passed parallel to where the telegraph cable parted in August, 1857. Our average speed has been about five knots. We were obliged to stop the screw engines in order to bring down to that speed, and, moreover, to reduce the paddle boiler power. Captain Anderson's ingenious mode of cleaning the ship's bottom, which he carried out last winter at Sheerness, has proved to have effected this very desirable object. Mr. Beckwith, the engineer, is now enabled to regulate and adjust her speed, and get more out of the ship than he could last year, when her bottom was one incrusted mass of mussels.

"Tuesday.—Another twenty-four hours of uninterrupted success. All day yesterday it was so calm that the masts of our convoy were reflected in the ocean, an unusual thing to see. A large shoal of porpoises gambolled about us for half an hour. A glorious sunset, and later, a crescent moon, which we hope to see in the brightness of her full, lighting our way into Trinity Bay before the days of this July shall have ended."

But the whole night did not pass away so tranquilly. By midnight the rain fell fast, and the wind blew fiercely, and then occurred the only real alarm of the voyage. The scene is thus described by Mr. Deane:

"All went on well until twenty minutes past twelve A.M., Greenwich time, when the first real shock was given to the success which has hitherto attended us, and this time we had real cause to be alarmed. A foul flake took place in the after tank. The engines were immediately turned astern, and the paying out of the cable stopped. We were all soon on deck, and learned that the running or paying-out part of the coil had caught three turns of the flake immediately under it, carried them into the eye of the coil, fouling the lay out, and hauling up one and a half turns from the outside, and five turns in the eye of the under flake. This was stopped, fortunately, before entering the paying out machinery. Stoppers of hemp also were put on near the V-wheel astern, and Mr. Canning gave orders to stand by to let go the buoy. This was not very cheering to hear, but his calm and collected manner gave us all confidence that his skill and experience would extricate the cable from the obvious danger in which it was placed. No fishing line was ever entangled worse than the rope was when thrust up in apparently hopeless knots from the eye of the coil to the deck. There at least five hundred feet of rope lay in this state, in the midst of thick rain and increasing wind. The cable crew set to work under their chief engineer's instructions to disentangle it. Mr. Halpin was there too, patiently following the bights as they showed themselves; the crew now passing them forward, now aft, until at last the character of the tangle was seen, and soon it became apparent that ere long the cable would be cleared. All this time Captain Anderson was at the taffrail anxiously watching the strain on the rope, which he could scarcely make out, the night was so dark, and endeavoring to keep it up and down, going on and reversing with paddle and screw. When one reflects for a moment upon the size of the ship, and the enormous mass she presents to the wind, the difficulty of keeping her stern, under the circumstances, over the cable, can be appreciated. The port paddle-wheel was disconnected; but shortly afterward there was a shift of wind, and the vessel canted the wrong way. Welcome voices were now heard passing the word aft from the tank that the bights were cleared, and to pay out. Then the huge stoppers were gently loosened, and at five minutes past two A.M., to the joy of all, we were once more discharging the cable. They veered it away in the tank to clear away the foul flake until three A.M., when the screw and paddle engines were slowed so as to reduce the speed of the ship to four and a half knots. During all this critical time there was an entire absence of noise and confusion. Every order was silently obeyed, and the cable men and crew worked with hearty good-will. Mr. Canning has had experience of foul flakes before, and showed that he knew what to do in the emergency. But what of the electrical condition of the cable during this period? Simply, that through its entire length it was perfect."

Thus, after three anxious hours, the danger was past, and the next morning the report of the ship is, "A fresh breeze from the southward, a dull gray sky, with occasional rain, and a moderate sea."

"Thursday.—There was a fresh breeze in the afternoon yesterday, increasing toward evening. It brought a heavy swell on the port quarter, which caused the ship to roll. The paying out from the after tank went on steadily. Two of the large buoys were lifted by derrick from the deck near the bows of the ship, and placed in position on the port and starboard side of the forward pick-up machinery, ready for letting go if necessary. The sun went down with an angry look, and the scud came rapidly from the eastward, the sea rising. A wind dead aft is not the best for cable laying, particularly if any accident should take place. By half-past eleven to-night we shall have exhausted the contents of the after tank, and the cable will then be paid out from the fore tank along the trough to the stern, the distance from the centre of the tank to the paying-out machinery being four hundred and ninety-four feet. Last night the swell was very heavy, to which the Great Eastern proved herself not insensible. Her rolling, like everything else appertaining to her, is done on a grand scale. We see the liveliness with which that operation is performed on board the Albany and Medway, and we are not at all disposed to be too critical in our observations on our own movements. The speed of the ship was kept at four and a half during the night—the slower the better, is the opinion of all on board—*festina lente*. We are consuming about one hundred tons a day of the seven thousand tons of coal which we had on board when we left Berehaven, and Mr. Beckwith, who has been engineer of the Great Eastern from her first voyage to the present moment, says her engines were never in better order; and their appearance and working do him and his able staff of assistant engineers the greatest credit.

"Friday.—Yesterday was a day of complete success, the paying out in every respect satisfactory. The wind still from the eastward, but inclined to draw to the northward, the sea entirely gone down. As Mr. Canning told us we should see the after tank emptied at eleven o'clock, ship's time, we were all collected there about ten o'clock, by which time the cable was down to the last flake. Next to having daylight for changing from the after to the fore tank, we could not have had a more favorable time—clear starlight, no wind, and a smooth sea. Looking down into the tank, the scene was highly picturesque. The cable-watch, whose figures were lighted up by the lamps suspended from above, slowly and cautiously lifted the turns of the coil to ease their path to the eye. As each found its way to the drum, the wooden floor of the tank showed itself, and then we saw more floor, and as its area increased the cable swept along its surface with a low, subdued noise, until, with a graceful curve, it mounted to the outlet, where it was soon to join a fresh supply ; and now we hear the word passed that they have arrived at the last turn, and the men who stood on the stages of the platform of the eye with the bight, watch the arrival of the cable and pass it up with tender caution, until it reaches the summit ; then it rushes down a wooden incline to meet the spliced rope, which had by this time come down along the trough leading from the forward tank. This operation was conducted with great skill by Mr. Canning and his experienced assistants, Messrs. Clifford and Temple. At eleven minutes past one A.M. (Greenwich time), the fresh rope was going over the stern, and the screw engines going ahead at thirteen minutes past one. A watch of four men is now stationed, fore and aft, all along the trough, which is illuminated by many lamps at short distances from each other. A lamp with a green light indicates the mile-mark as it comes up from the tank, and this signal is repeated until it reaches the stern, where it is recorded by the clerk who keeps the cable-log, in an office adjoining the paying-out machinery. A red lamp indicates danger. During the daytime red and blue flags are used. All through the night the sea was smooth as glass, and by this morning we saw that a sensible impression had been made on the contents of the fore tank. The ship begins to lighten at the bows, and by this time to-morrow will come up more as the cable passes out of the tank.

"Saturday.—Yesterday was our seventh day of paying out cable, and so far we have been more fortunate than the expedition of last year. During the same period of 1865, two faults had occurred—one on the twenty-fourth July, the other on the twenty-ninth—causing a detention of fifty-six hours. At three P.M. we were half-way, and passed where the Atlantic Cable of 1858 parted twice, on the twenty-sixth and twenty-eighth of June—sad memories to many! We feel, however, that every hour is increasing our chance of effecting this great work. 'I believe we shall do it this time, Jack,' I heard one of our crew say to another last night. 'I believe so too, Bill,' was the reply; 'and if we don't, we deserve to do it, and that's all.' It blew very hard from two o'clock yesterday, up to 10 P.M., by which time the wind gradually found its way from south-west to north-west, which is right ahead, just what we want for cable-laying. The Terrible and the two other ships plunged into the very heavy sea which the southwester raised, and we made up our minds, from what we saw, that the Great Eastern is the right ship to be in, in a gale of wind. During the night heavy showers of rain. This morning the sea was comparatively smooth, and the sky showed welcome patches of bright blue. If all goes well, we shall be up to-morrow evening at the place where last year's cable parted. A couple of days would bring us to shallower water, and then we may fairly look out for our 'Heart's Content.' Messages come from England, with the news, regularly and speedily—excellent practice for the clerks on shore and on board ship—great comfort to us, and the best evidence to those who will read this journal, of the great

VICTORY AT LAST.   329

fact that, up to this time, the cable is doing its electric work efficiently."

The interest of the voyage was greatly increased by the news daily received from Europe. Though in the middle of the Atlantic, they were still joined with the Old World, and messages came to the "Great Eastern Telegraph" as regularly as to the Times in London; reporting the quotations of the Stock Exchange, the debates in Parliament, and all the news of home. But what was far more exciting, was the tidings of the great events transpiring on the Continent. While the expedition had been preparing in England, a war had broken out of tremendous magnitude. Austria, Prussia, and Italy had rushed into the field. Armies, such as had not met since the fatal day of Leipsic, stood in battle array, and the thunder of war was echoing and reëchoing among the mountains of Bohemia. Amid these convulsions the fleet set sail; but it was still linked with the nations which it left behind, and received tidings from day to day. What great events were thus heralded to them in mid-ocean may be seen by a few items gleaned from the numerous despatches:

"Saturday evening, July 14th.—General Cialdini is moving upon Rovigo with an army of one hundred thousand men and two hundred guns. The Austrians have evacuated the whole country between the Mincio and Adige."

A day or two later:

"Cialdini has occupied Padua, twenty-three miles from Venice, on the railway connecting that city with the Quadrilateral, and the Austrians are shut up in Venice."

"Tuesday, 17th.—Prussians had successful engagement before Olmütz yesterday; captured six guns. Further fighting expected to-day. Austrians withdrawing from Moldavia toward Vienna."——"Conflict between Prussians and Federals. Prussians completely victorious. Federals evacuating Frankfort, and Prussians marching there."

"Thursday, 19th.—Prussians repeating victories, and gaining adhesions from small States. The main army within fifty miles of Vienna—have cut the railway to Vienna. Austrian army between Prussians and Vienna, under Archduke, one hundred and sixty thousand men. Money and archives removed from Vienna to Comorn."

"20th.—Frankfort occupied by the Prussians, who are marching on Vienna. Yesterday, Italian fleet, consisting of iron-clad vessels and several steamers, opened attack on Island of Lissa on the coast of Dalmatia—result not known."

The next day it is reported thus:

"Severe naval engagement off Lissa. Austrians claim the victory. Sunk one Italian iron-clad, run down another, blew up a third."

"July 21st.—Prussians crossed river; march near Holitzon, Hungary. Austria accepted proposal of armistice. Prussia will abstain from hostilities for five days, during which Austria will have to notify acceptance of preliminaries. A long letter published from the King of Prussia to the Queen, giving account of battle of Königgrätz."

The interest excited by such news may be imagined, coming while the events were yet fresh. Twice

VICTORY AT LAST.   331

a day was the bulletin set up on the deck, and was surrounded by an eager crowd reading what had transpired on the Continent but a few hours before. Nor was the intelligence confined to the Great Eastern. By an arrangement of signals, more complete than ever was used in a squadron before, the news was telegraphed to the convoy. All the ships had been furnished with experienced signal-men by the Admiralty. The system adopted was that known as Colomb's Flash Signals, by which, even in the darkest night, messages could easily be flashed to a distance of several miles. Thus all the ships were supplied with news twice a day, and the great military events in Europe were discussed in every cabin as eagerly as in the clubs of London. Again Deane's Diary reports:

"Sunday, July 22d.—Still success to record. A bright clear day, with a fresh and invigorating breeze from the north-west. Cable going out with unerring smoothness, at the rate of six miles an hour. There has been great improvement in the insulation. This remarkable improvement is attributable to the greatly decreased temperature of, and pressure on, the cable in the sea. This is a very satisfactory result to Mr. Willoughby Smith. Signals, too, come every hour more distinctly. This morning the breeze freshened. We are now about thirty miles to the southward of the place where the cable parted on the second of August, 1865, having then paid out one thousand two hundred and thirteen miles. Captain Anderson read divine service in the dining saloon.

"Monday.—Between six and seven P.M. yesterday, we

passed over the deepest part of our course. There was no additional strain on the dynamometer, which indicated from ten to fourteen hundred, the cable going out with its accustomed regularity. The wind still fresh from the north-west. During the night it went round to the southwest, and this morning there is a long roll from the southward.

"At forty-six minutes past eleven A.M., Mr. Cyrus Field sent a message to Valentia, requesting Mr. Glass to obtain the latest news from Egypt, India, and China, and other distant countries, so that on our arrival at Heart's Content we shall be able to transmit it to the principal cities of the United States. In just eight minutes he had a reply in these words: 'Your message received, and is in London by this.' Outside the telegraph room there is a placard put up, on which is posted the news shortly after its arrival, and groups of the crew may be seen reading it, just as we see a crowd at a newspaper office in London. Mr. Dudley, the artist, has made a very spirited sketch of 'Jack' reading the morning news, for he is supplied with the latest intelligence from the seat of war twice a day !* How he will grumble when he gets ashore ! He is not going to pay a pound a word for news, but his newspapers will supply it to him, and he does not know or care what it costs. But what a sum has been spent in Atlantic telegraphs ! It cannot now fall short of two millions and a half of pounds, or over twelve millions of dollars. More millions will be found if it shall be practically proved that America can permanently talk to England, and through her to the eastern hemisphere, and England to America by this ocean wire. At a quarter to twelve to-day but

---

* Mr. Dudley made a number of sketches for Mr. Field, with several large paintings, which have furnished the illustrations for this volume.

two hundred and fifteen miles of cable remained to be paid out of the fore tank. To-morrow night we hope to see it empty—then, for a small supply from the main tank, and then——but, hopeful though we are, let us not anticipate.

"Tuesday.—Breakfast at eight. Lunch at one. Dinner at six. Tea at eight. Five hundred and two souls who live on board this huge ship following their prescribed occupations. Cable going out merrily. · Electrical tests and signals perfect, and this is the history of what has taken place from noon yesterday to noon to-day. May we have three days more of such delightful monotony! It rained very hard during yesterday evening, and as we approach the Banks of Newfoundland we get thick and hazy weather."

The latter part of the voyage did not fulfil in all respects the promise of the first. The bright skies were gone; and instead perpetual fog hung over the water, while often the clouds poured down their floods. Thus the diary continues:

"Wednesday.—Fog and thick rain—just the weather to expect on approaching the Banks of Newfoundland. The convoy keep their position, and though sometimes the fog hides the ships from our view, yet we know where they are by their signal-whistles—two from the Terrible, three from the Medway, and four from the Albany, which are replied to by the prolonged single shriek from our whistle. At fifty-two minutes past one, Greenwich time (ship's time, forty-five minutes past ten P.M., last night), the fore tank being nearly empty, preparations were made for passing the bight of the cable into the main tank. At fifteen minutes past two all the jockey-wheels of the paying-out machinery were up, and the brakes released. Twenty-three minutes past two the bight was passed steadily and cautiously by the cable hands outside of the trough to the main tank, and at thirty-five minutes past two the splice went over the stern in 1542.8 fathoms. By arrangement with Sir James Hope, the admiral of the North-American station, who has received instructions from the Admiralty to give the present expedition every assistance in his power, a frigate or sloop will be placed in longitude 48°, 25′, 52″, which is just thirty miles from the entrance of Trinity Bay, and sixty from Heart's Content. She will probably hang on by a kedge in that position, which shows the 'fair way' right up the bay; and if it be clear, we ought to see her about daybreak on Friday morning. The fog was very thick this morning, but occasionally lifts; as long as the wind is from south-west we cannot expect clear weather."

As the week drew on, it was evident that they were approaching the end of their voyage. By Thursday they had passed the great depths of the Atlantic, and were off soundings. Besides their daily observations, there were many signs, well known to mariners, that they were near the coast. There were the sea-birds, and they could almost snuff the smell of the land, such as once greeted the sharp senses of Columbus, and made him sure that he was floating to some undiscovered shore. Captain Anderson had timed his departure so that he should approach the American coast at the full moon; but for the last two or three nights, as the round orb rose behind them, banks of cloud hung so heavily upon the water, that the moonlight only gleamed faintly through the vaporous air, and the fleet seemed like the phantom ships of the Ancient Mariner, drifting on through fog and mist.

"Thursday.—All day yesterday it was as 'thick as mustard.' We have had now forty-eight hours of fog. Though it lifted a little this morning, at five A.M., it still looks like more of it. Captain Anderson signalled to the Albany, at fifteen minutes past ten last night, to start at daybreak, and proceed to discover the station ship, and report us at hand. Should she fail to find her, then to try and make the land and guide us up Trinity Bay. Another signal was sent at half-past twelve to the effect that the Terrible and Medway would be sent ahead to meet the Albany and establish a line to lead us in even with a fog. The Albany started at half-past three. At forty-five minutes past four, Greenwich time, the cable engineer in charge took one weight off each brake of the paying-out machinery. At forty minutes past seven all weights taken off, the assumed depth being three hundred fathoms. The indicated strain on the dynamometer gradually decreasing. Speed of ship five knots. We are going to try and pick up the cable of 1865 in two thousand five hundred fathoms (and we mean to succeed too); therefore should the cable we are now paying out part, it can be understood how easy it would be to raise it from a depth of three hundred fathoms. At fifty-five minutes past eight we signalled to the Terrible to sound, and received a reply, one hundred and sixty fathoms. At half-past eleven we informed her that when at the buoy off Heart's Content she should have her paddle-box boat and two cutters ready to be alongside immediately, for holding the bight of the cable during the splice and laying the shore end. The Medway was told at the same time to prepare two five-inch ropes, and two large mushroom anchors, with fifty fathoms of chain, for anchoring during the splice in one hundred and seventy fathoms of water, and we intimated to her that when inside of Trinity Bay we should signal for two boats to take hands on board her for shore end. News of to-day, telegram from Mr. Glass in reply to one from Mr. Canning: I congratulate you all most sincerely on your arrival in one hundred and thirty fathoms. I hope nothing will interfere to mar the hitherto brilliant success, and that the cable will be landed to-morrow.'"

As the voyage is about to end, we give the distances run from day to day, which show a remarkable uniformity of progress:

	Distance Run.	Cable Paid Out.
Saturday, fourteenth,	108	115
Sunday, fifteenth,	128	139
Monday, sixteenth,	115	137
Tuesday, seventeenth,	117	138
Wednesday, eighteenth,	104	125
Thursday, nineteenth,	112	129
Friday, twentieth,	117	127
Saturday, twenty-first,	121	136
Sunday, twenty-second,	123	133
Monday, twenty-third,	121	138
Tuesday, twenty-fourth,	120	135
Wednesday, twenty-fifth,	119	130
Thursday, twenty-sixth,	128	134
Friday, twenty-seventh,	100	104

From this it appears that the speed of the ship was exactly according to the running time fixed before she left England. On the last voyage it was thought that she had once or twice run too fast, and thus exposed the cable to danger. It was, therefore, decided to go slowly but surely. Holding her back to this moderate pace, her average speed, from the time the splice was made till they saw land, was a little less than five nautical miles an hour, while the cable was paid out at an average of not quite five and a half miles. Thus the total slack was about eleven per cent, showing that the cable was laid almost in a straight line, allowing for the swells and hollows in the bottom of the sea.

"Friday, July 27th.—Shortly after two P.M., yesterday, two ships, which were soon made out to be steamers, were seen to the westward; and the Terrible, steaming on ahead, in about an hour signalled to us that H.M.S. Niger was one of them, accompanied by the Albany. The Niger, Captain Bruce, sent a boat to the Terrible as soon as he came up with her. The Albany shortly afterward took up her position on our starboard quarter, and signalled that she spoke the Niger at noon, bearing E. by N., and that the Lily was anchored at the station in the entrance of Trinity Bay, as arranged with the Admiral. The Albany also reported that she had passed an iceberg about sixty feet high. At twenty minutes after four P.M., the Niger came on our port side, quite close, and Captain Bruce, sending the crew to the rigging and manning the yards, gave us three cheers, which were heartily returned by the Great Eastern. She then steamed ahead toward Trinity Bay. The Albany was signalled to go on immediately to Heart's Content, clear the northeast side of the harbor of shipping, and place a boat with a red flag for Captain Anderson to steer to, for anchorage. Just before dinner we saw on the southern horizon, distant about ten miles, an iceberg, probably the one which the Albany met with. It was apparently about fifty or sixty feet in height. The fog came on very thick about eight P.M., and between that and ten we were constantly exchanging guns and burning blue lights with the Terrible, which, with the Niger, went in search of the Lily, station ship. The Terrible being signalled to come up and take her position, informed us they had made the Lily out, and that she bore then about E. N. E. distant four miles. Later in the night Captain Commerill said that if Captain Anderson would stop the Great Eastern, he would send the surveyor Mr. Robinson, R. N., who came out in the Niger, on board of us, and about three the engines were slowed, and the Terrible shortly afterwards came alongside with that officer. Catalina light, at the entrance of Trinity Bay, had been made out three hours before this, and the loom of the coast had also been seen. Fog still prevailing! According to Mr. Robinson's account, if they got one clear day in seven at the entrance of Trinity Bay, they considered themselves fortunate. Here we are now (six A.M.), within ten miles of Heart's Content, and we can scarcely see more than a ship's length. The Niger, however, is ahead, and her repeated guns tell us where we are with accuracy. Good fortune follows us, and scarcely has eight o'clock arrived when the massive curtain of fog raises itself gradually from both shores of Trinity Bay, disclosing to us the entrance of Heart's Content, the Albany making

VICTORY AT LAST. 339

for the harbor, the Margaretta Stevenson, surveying vessel, steaming out to meet us, the preärranged pathway all marked with buoys by Mr. J. H. Kerr, R. N., and a whole fleet of fishing boats fishing at the entrance.

"We could now plainly see that Heart's Content, so far as its capabilities permitted, was prepared to welcome us. The British and American flags floated from the church and telegraph station and other buildings. We had dressed ship, fired a salute, and given three cheers, and Captain Commerill of H.M.S. Terrible was soon on board to congratulate us on our success. At nine o'clock, ship's time, just as we had cut the cable and made arrangements for the Medway to lay the shore end, a message arrived giving us the concluding words of a leader in this morning's Times: 'It is a great work, a glory to our age and nation, and the men who have achieved it deserve to be honored among the benefactors of their race.'—'Treaty of peace signed between Prussia and Austria!' It was now time for the chief engineer, Mr. Canning, to make the necessary preparations for splicing on board the Medway. Accompanied by Mr. Gooch, M.P., Mr. Clifford, Mr. Willoughby Smith, and Messrs. Temple and Deane, he went on board, the Terrible and Niger having sent their paddle-box boats and cutters to assist. Shortly afterward the Great Eastern steamed into the harbor and anchored on the north-east side, and was quickly surrounded by boats laden with visitors. Mr. Cyrus Field had come on shore before the Great Eastern had left the offing, with a view of telegraphing to St. John's to hire a vessel to repair the cable unhappily broken between Cape Ray, in Newfoundland, and Cape North, in Breton Island. Before a couple of hours the shore end will be landed, and it is impossible to conceive a finer day for effecting this our final operation. Even here, people can scarcely realize the fact that the Atlantic Telegraph Cable has been laid. To-morrow, however, Heart's Content * will awaken to the fact that it is a highly favored place in the world's esteem, the western landing-place of that marvel of electric communication with the Eastern hemisphere, which is now happily, and we hope finally, established."

This simple record, so modestly termed the Diary of the Expedition, tells the story of this memorable voyage in a way that needs no embellishment. But if from the ship's deck we transfer ourselves to the shore, we may get a new impression of the closing scene. We can well believe the sensation of wonder and almost of awe, on the morning when the ships entered that little harbor of Newfoundland. In England the

---

* The little harbor that bears this gentle name, is a sheltered nook where ships may ride at anchor, safe from the storms of the ocean. It is but an inlet from the great arm of the sea known as Trinity Bay, which is sixty or seventy miles long, and twenty miles broad. On the beach is a small village of some sixty houses, most of which are the humble dwellings of those hardy men who vex the northern seas with their fisheries. The place was never heard of outside of Newfoundland till 1864, when Mr. Field, sailing up Trinity Bay in the surveying steamer Margaretta Stevenson, Captain Orlebar, R. N., in search of a place for the landing of the ocean cable, fixed upon this secluded spot. The old landing of 1858 was at the Bay of Bull's Arm, at the head of Trinity Bay, twenty miles above. Heart's Content was chosen now because its waters are still and deep, so that a cable skirting the north side of the Banks of Newfoundland can be brought in deep water almost till it touches the shore. All around the land rises to pine-crested heights ; and here the telegraphic fleet, after its memorable voyage, lay in quiet, under the shadow of the encircling hills.

progress of the expedition was known from day to day, but on this side of the ocean all was uncertainty. Some had gone to Heart's Content, hoping to witness the arrival of the fleet, but not so many as the year before, for the memory of the last failure was too fresh, and they feared another disappointment. But still a faithful few were there, who kept their daily watch. The correspondents of the American papers report only a long and anxious suspense, till the morning when the first ship was seen in the offing. As they looked toward her, she came nearer—and see, there is another and another! And now the hull of the Great Eastern loomed up all glorious in the morning sky. They were coming! Instantly all was wild excitement on shore. Boats put off to row toward the fleet. The Albany was the first to round the point and enter the bay. The Terrible was close behind. The Medway stopped an hour or two to join on the heavy shore end, while the Great Eastern, gliding calmly in as if she had done nothing remarkable, dropped her anchor in front of the telegraph house, having trailed behind her a chain of two thousand miles, to bind the Old World to the New.

That same afternoon, as soon as the shore end was landed, Captain Anderson and the officers of the fleet went in a body to the little church in Heart's Content, to render thanks for the success of the expedition. A sermon was preached on the text, " There shall be no more sea," and all joined in the sublime prayers and thanksgivings of the Church of England. Thus the voyage ended as it began. It left the shores of Ireland with prayers wafted after it as a benediction. And now, safely landed on the shores of the New World, this gallant company, like Columbus and his companions, made it their first thought to render homage to the Being who had borne them safely across the deep.

But though their voyage was ended, there was still a work to be done. Having crossed the Atlantic, the first thing was to open communication with the cities of the United States. And now Mr. Field was extremely mortified to find that there was a large gap in the line this side of the ocean. His first question to the Superintendent, who came out in a boat to meet him, was in regard to the cable across the Gulf of St. Lawrence, which had been interrupted the year before; and it was a bitter pang to hear that it lay still broken, so that a message which came from Ireland in a moment of time, was delayed twenty-four hours in its way to New York. Of course the public grew impatient, and there were many sneers at the want of foresight which had failed to provide against such a contingency; and, as he was the one chiefly known in connection with the enterprise, these reproaches fell upon him. He did not tell the public, what was the truth, that he had anticipated this very trouble long ago, and entreated his associates to be prepared for it.

Months before he left for England, he urged upon the Company in New York the necessity of rebuilding their lines in Newfoundland, which had been standing over ten years, and of repairing the old cable, and also laying a new one across the Gulf of St. Lawrence. But this would cost a large sum of money, and as their faith and purses had been sorely tried by repeated disasters, they were not willing to spend more in the uncertainty of the future. They wished to see the result of this new expedition, before advancing further capital. We do not blame them, but only mention the fact to show that Mr. Field had foreseen this very thing, and endeavored to guard against it.

But regrets were idle. What could he do to repair the injury? "Is there a steamer," he asked, "to be had in these waters?" "The Bloodhound is at St. John's." "Telegraph instantly to charter her to go around to the Gulf of St. Lawrence, and fish up the old cable and repair it. But that may take several weeks. Is there nothing else that can serve in the mean time to carry despatches across the Gulf?" "There is a little steamer, called the Dauntless." "Well, telegraph for her too. Secure her at all hazards; only see that the work is done." All this was the work of a few minutes. The answers came back quickly, and in a day or two came the steamers themselves. The arrangement was immediately carried out. The Dauntless took her place in the Gulf, where she made her regular trips from Port au Basque, in Newfoundland, to Aspee Bay, in Cape Breton, keeping up daily communication with the States. The Bloodhound, which had a more difficult task, first took on board eleven miles of cable from the Great Eastern, to repair that which was broken. The expedition was put in charge of Mr. A. M. Mackay, the indefatigable Superintendent of the Company in Newfoundland, who had had the care of their lines for ten years. He sailed for Aspee Bay, and made short work of the business, dragging the Gulf and raising the cable, which he found had been broken by an anchor, in water seventy fathoms deep, a few miles from shore. This was spliced out with a portion of the new cable, and the whole was as perfect as ever, giving a fresh proof that cables well made are likely to be permanent, if not indestructible.

Meanwhile, owing to this interruption of the cable across the Gulf of St. Lawrence, the news of the success of the expedition, which reached Newfoundland on Friday, the twenty-seventh, did not reach New York till the twenty-ninth. It was early Sunday morning, before the Sabbath bells had rung their call to prayer, that the tidings came. The first announcement was brief: "Heart's Content, July 27.—We arrived here at nine o'clock this morning. All well. Thank God, the cable is laid, and is in perfect working order. CYRUS W. FIELD."

Soon followed the despatch to the Associated Press, giving the details of the voyage, and ending with a just tribute to the skill and devotion of all who had contributed to its success. Said Mr. Field: "I cannot find words suitable to convey my admiration for the men who have so ably conducted the nautical, engineering, and electrical departments of this enterprise, amidst difficulties which must be seen to be appreciated. In fact, all on board of the telegraph fleet, and all connected with the enterprise, have done their best to have the cable made and laid in a perfect condition; and He who rules the winds and the waves has crowned their united efforts with perfect success."

Other despatches followed in quick succession, giving the latest events of the war in Europe, which startled the public just reading news a fortnight old. All this confirmed the great triumph, and filled every heart with wonder and gratitude on the Sunday morning, as they went again to the little church and rendered thanks to Him who is Lord of the earth and sea.

While the Great Eastern was lying in the harbor of Heart's Content, she was overrun with visitors. The news of her arrival had spread over the island, and from far and near the people flocked to see her. Over the hills they came on foot and on horseback, and in wagons and carts of every description; and from along the shore in boats and fishing-smacks, and sloops and schooners. They came from the most remote parts of the island—a distance of three hundred miles—and even from the province of New Brunswick. Several parties made the excursion in steamers from St. John's. The wondering country folk climbed up the sides of the ship, and wandered for hours through its spacious rooms and long passages. All were welcomed with hearty sailor courtesy.

As soon as communication was opened with New York, and other cities, congratulations poured in from every quarter. Friendly messages were exchanged—as eight years before—between the sovereign of England and the head of the Great Republic. The President also, and Mr. Seward, Secretary of State, sent their congratulations to Mr. Field—greetings that were repeated from the most distant States. Among others was a message from San Francisco, which was put into his hand almost at the same moment with one from M. de Lesseps, dated at Alexandria in Egypt! What a meeting and mingling of voices was this, when a winged salutation flying over the tops of the Rocky Mountains, reached the same ear with a message which had been whispered along the Mediterranean and under the Atlantic: when the farthest East touched the farthest West—the most ancient of kingdoms answering to the new-born empire of the Pacific.

CHAPTER XVII.

RECOVERY OF THE LOST CABLE.

THOUGH the Great Eastern was still lying in the little harbor of Heart's Content, casting her mighty shadow on its tranquil waters, she was not "content" with her amazing victory, but sighed for another greater still. Though she had done enough to be laid up for a year, still she had one more test of her prowess—to recover the cable of 1865, which had been lost in the middle of the Atlantic. So eager were all for this second trial of their strength, that in less than five days two of the ships—the Albany and the Terrible—the vanguard of the telegraphic fleet, were on their way back to mid-ocean. Though it was only Friday, the 27th of July, that they reached land, they left early Wednesday morning, the first day of August. The Great Eastern was detained a week longer. She had to lay in immense supplies of coal. Anticipating this want, six ships had been despatched from Cardiff, in Wales, weeks before, to await the arrival of the fleet. One of these foundered at sea; the others arrived out safely, and hardly had the Great Eastern cast anchor before they were alongside, ready to fill her bunkers.

So ample was the provision, that, when she went to sea a few days after, she had nearly eight thousand tons of coal on board.

At the same time she had to receive some six hundred miles of the cable of 1865, which had been shipped from England in the Medway. The latter was now brought alongside, and the whole was transferred into the main tank of the Great Eastern, from which it was to be paid out in case the lost end were recovered.

At length all these preparations were completed, and on Thursday, the 9th of August, the Great Eastern and the Medway put to sea. The Governor of Newfoundland, who had come around from St. John's and been received with the honors due his rank, accompanied them in the Lily down the broad expanse of Trinity Bay, and then bore away for St. John's while the Great Eastern and Medway kept on their course to join their companions in the middle of the Atlantic. They had a little over six hundred miles to run to the "fishing ground," and made it in three days. On Sunday noon they came in sight of the appointed rendezvous, and soon with glasses made out the Albany and the Terrible, which had arrived a week before and placed buoys to mark the line of the cable, and then, like giant sea-birds with folded wings, sat watching their prey. The sea was running high, so that boats could not come off, but the Albany signalled that she had not toiled for nothing; that she had once hooked the cable, but lost it in rough weather. The history of this first attempt, though brief, was cheering.

When the Albany left Heart's Content, Captain Moriarty went in her. He had been in the Great Eastern the year before, and saw where the cable went down, and had had his eye on the spot ever since. He claimed, with Captain Anderson, that he could go straight to it and place the ship within half a mile of where it disappeared. At this old sailors shook their heads, and said, "They'd like to see him do it;" "No man could come within two or three miles of any given place in the ocean." Yet the result proved the exactness of his observations. With unerring eye he went straight to the spot, and set his buoys as exactly as a fisherman sets his nets.

In the Albany, also, had gone Mr. Temple, of Mr. Canning's staff. The ship had been fitted up with a complete set of buoys and apparatus for grappling; and he was full of ambition to recover the cable before the Great Eastern should come up. In this he had nearly proved successful. They had caught it once, and raised it a few hundred fathoms from the bottom, and buoyed it, but rough weather came on and tore away the buoy, so that the cable went down again, carrying two miles of rope.

This was a disappointment, but still, as their first attempt was only a "feeler," the result was encouraging. It showed that they had found the right place; that the cable was there; that it had not run away nor been floated off by those under-currents that exist in the imagination of some wise men of the sea; nor that it was so imbedded in the ooze of the deep as to be beyond reach or recovery. All this was cheering, but as it promised to be a more difficult job than they had supposed, they were glad when the Great Eastern hove in sight that Sunday noon.

The next morning Captain Moriarty and Mr. Temple came on board, and after reporting their experience, the chief officers of the Expedition held a council of war before opening the campaign. The fleet was all together, the weather was favorable, and it was determined at once to proceed to business.

As the attempt is now to be renewed on a grand scale, the reader may wish some further details of the means employed to insure success. As nothing in this whole enterprise has excited such astonishment, nothing merits a more careful history. When it was first proposed to drag the bottom of the Atlantic for a cable lost in waters two and a half miles deep, the project was so daring that it seemed to be almost a war of the Titans upon the gods. Yet never was anything undertaken less in the spirit of reckless desperation. The cable was recovered, as a city is taken by siege— by slow approaches, and the sure and inevitable result of mathematical calculation. Every point was studied beforehand—the position of the broken end, the depth of the ocean, the length of rope needed to reach the bottom, and the strength required to lift the enormous weight. To find the place was a simple question of nautical astronomy—a calculation of latitude and longitude. It seemed providential that, when the cable broke on the second of August, 1865, it was a few minutes after noon; the sun was shining brightly, and they had just taken a perfect observation. This made it much easier to go back to the place again. The waters were very deep, but that they could touch bottom, and even grapple the cable, was proved by the experiments of the year before. But could any power be applied which should lift it without breaking, and bring it safely on board? This was a simple question of mechanics. Prof. Thomson had made a calculation that in raising the cable from a depth of two and a half miles, there would be about ten miles of its length suspended in the water. Of course, it was a very nice matter to graduate the strain so as not to break the cable. For this it had been suggested that two or three ships should grapple it at once, and lifting it together, ease the strain on any one point—a method of meeting the danger that was finally adopted with success.

With such preparations, let us see how all this science and seamanship and engineering are applied. The ships are now all together in the middle of the

Atlantic. The first point is achieved. They have found the place where the broken cable lies—they have laid their hands on the bottom of the ocean and "felt of it," and know that it is there. The next thing is to draw a line over it, to mark its course, for in fogs and dark nights it cannot be traced by observations. The watery line is therefore marked by a series of buoys a few miles apart, which are held in position by heavy mushroom-anchors, let down to the bottom by a huge buoy-rope, which is fastened at the top by a heavy chain. Each buoy is numbered, and has on the top a long staff with a flag, and a black ball over it, which can be seen at a distance. Thus the ships, ranging around in a circuit of many miles, can keep in sight this chain of sentinels. The buoy which marks the spot where they wish to grapple has also a lantern placed upon it at night, which gleams afar upon the ocean. Having thus fixed their bearings, the Great Eastern stands off, north or south according to the wind or current, three or four miles from where the cable lies, and then, casting over the grapnel, drifts slowly down upon the line, as ships going into action reef their sails, and drift under the enemy's guns.

The "fishing-tackle" is on a gigantic scale. The "hooks," or grapnels, are huge weapons armed with teeth, like Titanic harpoons to be plunged into this submarine monster. The "fishing-line" is a rope six and a half inches round, and made of twisted hemp and iron, consisting of forty-nine galvanized wires, each bound with manilla, the whole capable of bearing a strain of thirty tons. Of this heavy rope there are twenty miles on board the ships, the Albany carrying five, and the Great Eastern and the Medway seven and a half miles each. Of course it is not the easiest thing in the world to handle such a rope. But it is paid out by machinery, passing over a drum; and the engine works so smoothly, that it runs out as easily as ever a fisherman's line was reeled off into the sea. As it goes out freely, the strain increases every moment. The rope is so ponderous, that the weight mounts up very fast, so that by the time it is two thousand fathoms down, the strain is equal to six or seven tons. The tension of course is very great, and not unattended with danger. What if the rope should break? If it should snap on board, it would go into the sea like a cannon-shot. Such was the tension on the long line, that once when the splice between the grapnel-rope and the buoy-rope "drew," the end passed along the wheels with terrific velocity, and flying in the air over the bow, plunged into the sea. But the rope is well made, and holds firmly an enormous weight. It takes about two hours for the grapnel to reach the bottom, but they can tell when it strikes. The strain eases up, and then, as the ship drifts, it is easy to see that it is not dragging through the water, but over the ground.

"I often went to the bow," says Mr. Field, "and sat on the rope, and could tell by the quiver that the grapnel was dragging on the bottom two miles under us."

And thus, with its fishing line set, the great ship moves slowly down over where the cable lies. As the grapnel drags on the bottom, one of the engineer's staff stands at the dynamometer to watch for the moment of increasing strain. A few hours pass, and the index rises to eight, ten, or twelve tons, sure token that there is something at the end of the line—it may be the lost cable, or a sunken mast or spar, the fragment of a wreck that went down in a storm that swept the Atlantic a hundred years ago. And now the engine is set in motion to haul in. As the rope comes up, it passes over a five-foot drum, every revolution bringing up three fathoms. Thus it takes some hours to haul in over two miles' length, perhaps at last to find nothing at the end!

Success in hooking the cable depends on the accuracy of their observations. These were sometimes verified in a remarkable manner. When the nights were very dark and thick with fog, so that they could not see the stars above nor their lights on the ocean, they had to go almost by the sense of feeling. Yet so exactly had they taken their bearings, that they could almost grope over the ground with their hands. A singular proof of this was given one night, when, just as the line began to quiver, showing that the cable had been hooked, one of the buoys—which had not been seen in the darkness—thumped against the side of the ship. So exactly had it been placed over the prescribed line, that the ship struck the buoy just as the grapnel struck the cable! The accident, which startled them at first, when it occurred in the gloom of night, furnished the strongest proof of the accuracy of their observations; and the officers were very proud of it, as they well might be, as a victory in nautical astronomy!

These different experiments revealed some secrets of the ocean. Its bottom proved to be generally ooze, a soft slime. When the rope went down, one or two hundred fathoms at the end would trail on the sea floor; and when it came up, this was found coated with mud, "very fine and soft like putty, and full of minute shells." But it was not *all* ooze at the bottom of the sea, even on this telegraphic plateau. There were hidden rocks—perhaps not cliffs and ledges, but at least scattered boulders, lying on that mighty plain. Sometimes the strain on the dynamometer would suddenly go up three or four tons, and then back again, as if the grapnel had been caught and broken away. Once it came up with two of its hooks bent, as if it had come in contact with a huge rock. At one time it brought up in the mud a small stone half the size of an almond; and at another a fragment as large as a brick. This was a piece of granite.

Friday, August 17th, was a memorable day in the expedition, for the cable was not only caught, but brought to the surface, where it was in full sight of the whole ship, and yet finally escaped. The day before the line had been cast over, at about two o'clock, and struck the ground a little before five. After dragging a couple of hours, the increasing strain showed that they had grappled the prize, and they began to haul in, but soon ceased, and held on till morning. Then the engine was set in motion again, and slowly but steadily the ponderous rope came up from the deep. By half-past ten o'clock, Friday morning, twenty-three hundred fathoms had come on board, and but fifteen or twenty remained. Then was the critical moment, and they paused before giving a last pull. Such was the eagerness of all, that the diver of the ship, Clark, begged to be allowed to plunge down twenty fathoms, to lay his hand on the prize, and be sure that it was there. But patience yet a few minutes! A few more strokes of the engine, and the sea-serpent shows himself—a long black snake with a white belly. "On the appearance of the cable," says Deane, in his Diary of the Expedition, "we were all struck with the fact that one half of it was covered with ooze, staining it a muddy white, while the other half was in just the state in which it left the tank, with its tarred surface and strands unchanged, which showed that it lay in the sand only half embedded. The strain on the cable

RECOVERY OF THE LOST CABLE.

gave it a twist, and it looked as if it had been painted spirally black and white. This disposes of the oft-repeated assertion, that we should not be able to pull it up from the bottom, because it would be embedded in the ooze."

The appearance of the cable woke a tremendous hurrah from all on board. They cheered as English sailors are apt to cheer when the flag of an enemy is struck in battle. But their exultation came too soon. The strain on the cable was already mounting up to a dangerous point. Capt. Anderson and Mr. Canning were standing on the bow, and saw that the strands were going. They hastened men to its relief, but it was too late. Before they could put stoppers on it to hold it, it broke close to the grapnel, and sunk to the bottom. It had been in sight but just five minutes, and was gone. Instantly the feeling of exultation was turned to one of disappointment, and almost of rage, at the treacherous monster, that lifted up its snaky head from the sea, as if to mock its captors, and instantly dived to the silence and darkness below.

It was a cruel disappointment. Yet when they came to think soberly, there was no cause for despair, but rather for new confidence and hope. They had proved what they could do. But this detained them in the middle of the Atlantic for two weeks more.

It were idle to relate all the attempts of those two weeks. Every day brought its excitement. Whenever the grapnel caught, there was a suspense of many hours till it was brought on board. Several times they seemed on the point of success. Two days after that fatal Friday, on Sunday, August 19th, they caught the cable again, and brought it up within a thousand fathoms of the ship, and buoyed it. But Monday and Tuesday were too rough for work, and all their labor was in vain. Thus it was a constant battle with the elements. Sometimes the wind blew fiercely and drove them off their course. Sometimes the buoys broke adrift and had to be pursued and taken. Once or twice the boatswain's mate—a brave fellow, by the name of Thornton—was lowered in ropes over the bow of the ship and let down astride of a buoy; and though it spun round with him like a top, and his life was in danger, he held on and fastened a chain to it, by which it was swung on board.

The continued bad weather was the chief obstacle to success. Engineers had often grappled for cables in the North Sea and the Mediterranean; but there they could look for at least a few days when the sea would be at rest; but in the Atlantic it was impossible to calculate on good weather for twenty-four hours. For nearly four weeks that they were at sea, they had hardly four days of clear sunshine, without wind. Often the ocean was covered with a driving mist, and the ships, groping about like blind giants, kept blowing their shrill fog-trumpets, or firing guns, as signals to

RECOVERY OF THE LOST CABLE.

their companions that they were still there. Occasionally the sun shone out from the clouds, and gave them hope of better success. Once or twice we find in the private journal kept by Mr. Field, that it was "too calm;" there was not wind enough to drift the ship over the cable, so that the rope hung up and down from the bow, without dragging. One Sunday night he remembered, when the deep was hushed to a Sabbath stillness, the moon was shining brightly, and the ships floating over a "sea of glass," that suggested thoughts of a better world than this. Such times gave them fresh hopes, that might be disappointed on the morrow.

Once, however, the Albany, which had been off a few miles fishing on its own hook, suddenly appeared in the night, reporting a victory. All on board the Great Eastern were startled by the firing of guns. It was a little after midnight, and Mr. Field had gone below, worn out with the long suspense and anxiety, when Captain Anderson came rushing to his state-room with tidings that the cable was recovered! Both hurried on deck, and sure enough there was the Albany bearing down upon them, with her crew cheering in the wildest manner. The gallant Temple had conquered at last. But the next morning brought a fresh disappointment. They had indeed got hold of the cable, and brought its end on board, and afterward buoyed it, but when the Great Eastern went for it, it proved to be only a fragment some two miles long which had been broken off in one of the previous grapplings. However, they hauled it in, and kept it with pride, as their first trophy from the sea.

And so the days and weeks wore on; it was near the end of August, and still the prize was not taken. The courage of the men did not fail, but they were becoming worn out. The tension on their nerves of this long suspense was terrible. On Tuesday, August 28th, Mr. Temple was brought on board from the Albany, very ill. He was worn out with constant watching. Their resources, too, must in time be exhausted. On the evening of the 29th, Captain Commerill, of the Terrible, came on board, and reported the condition of his ship. He was one of the very best officers in the fleet, full of zeal, courage, and activity (having a good right hand in his first officer, Mr. Curtis), and always kept up a brave heart, even in the darkest days.* But his supplies were nearly

---

* Captain Anderson, in a letter published after the return to England, says: "Every officer and man of the expedition will have pleasant recollection of the cheerful zeal of Captain Commerill, V.C., and the officers of Her Majesty's ship Terrible. Captain Commerill frequently visited us in his boats, both in high seas and in calms, and his cheery way of saying, 'You'll do it yet,' 'What can I do?' and 'I'll do it,' was truly characteristic of him. The officers of the Terrible would do any thing for their captain, and entered heartily into the object of the voyage."

Such a tribute from one brave commander to another, is most honorable to both. In the same letter he recognizes, also, the services renexhausted. He had been out four weeks, and his coal was almost gone, and his men were on half rations. So he must leave the fishing ground for fresh supplies. It was a painful necessity. He mourned his fate, like a brave officer who is ordered away in the midst of a battle. But he submitted only with a determination to take in ammunition, and to come back in a few days to renew the struggle. Accordingly the Terrible left the same evening for St. John's.

At the same time it was decided that the three other ships should leave their present cruising ground, and try a new spot. As an old fisherman, who has cast his line in one place so often as to scare the fish away, sometimes has better luck in other waters, so they pro-

dered by the captains of the other ships: "I shall do but scant justice to Commanders Prowse and Batt, R.N., and Captains Eddington and Harris, Mercantile Marine, of the Medway and Albany, if I recall the three weeks spent upon the 'grappling ground,' where we were often separated by fog, gale, or darkness; yet whenever day dawned, or the fog cleared, there the squadron were to be seen, converging from different points towards the Mark Buoy, a small spot looking no bigger than a man's hat on the surface of the ocean. Unless all had concentrated their minds, and watched their ships and compasses night and day, no such beautiful illustration of nautical science could have been possible. The vessels of the squadron keeping always together, and commanded by men who knew the importance of keeping close enough to begin work whenever it was possible, and yet to avoid collision in fog, was of the greatest importance; and we owe much to that invaluable system of signalling by night and day, invented by Captain Colomb, R.N., which enabled us, even in dark nights, when two or three miles apart, to communicate or ascertain anything we desired."

posed to go east a hundred miles, to a place where the ocean was not quite so deep. Deane, in his Diary, calls it " the sixteen hundred fathom patch," but they found it nineteen hundred fathoms, or about two miles! So the next morning the Great Eastern, the Medway, and the Albany "pulled up stakes," that is, took in their buoys, and bore away to the east. In a few hours they reached the appointed rendezvous, and had set their buoys. The last day of August had come, and all seemed favorable for a final attempt. It was a clear day, with no wind. The sea had gone down, so that at noon it was a dead calm, as the three ships took their position in line, about two miles apart, ready to open their broadsides at once. The grapnel went over for *the thirtieth time.* Kind heaven favored its search, and at ten minutes before midnight it had found the cable, and fastened its teeth never to let go. Feeling something at the end of the rope, they began to haul in, but slowly at first, as an expert angler decoys a big fish by pulling gently on the line. Watching the dynamometer, they saw with delight the strain increase with every hundred fathoms. Up it went to eight, nine, ten tons! They had caught it, and no mistake. In about five hours they had drawn it up to within a thousand fathoms of the top of the water, where it hung suspended from the ship. But now came the critical point, for as it approached the surface the danger of breaking increased every moment. It required

RECOVERY OF THE LOST CABLE.

delicate handling. To make sure this time, the Great Eastern buoyed the cable, and moved off two or three miles to take a fresh grip in a new place; and having got a double hold, the Medway, which was two miles further to the west, was ordered to grapple for it also; and having caught it, to heave up with all force, till she should bring it on board or break it. This was done, and the old cable brought up within three hundred fathoms, and there broken. This at once lightened the strain and gave them an end to pull upon, whereupon the Great Eastern, having a lighter weight on the rope, drew up again, but still gently, watching the strain, lest the cable should break. These operations were very slow, and lasted many weary hours. It was a little before midnight on Friday night that the cable was caught, and it was after midnight Sunday morning that it was brought on board. How long that day seemed! Night turned to morning, and morning to noon, and noon to night again, and still the work was not done; still the great ship hung over the spot where its treasure was suspended in the deep. The sun went down, and the moon looked forth from driving clouds upon a scene such as the ocean never saw before. At a distance could be discerned the black hulls of the attendant ships, the Albany and the Medway. But why were they thus silent and motionless in the midst of the sea? Some mysterious errand brought them here, and as their boats approached with measured sweep, at this midnight hour, it seemed as if they came with muffled oars to an ocean burial. It was still calm, but the sea began to moan with unrest, as if troubled in its sleep. As midnight drew on, the interest gathered about the bows of the Great Eastern. The bulwarks were crowded with anxious watchers, peering into the darkness below. Still not a word was spoken. Not a voice was heard, save that of Captain Anderson, or Mr. Halpin, or Mr. Canning, giving orders. As it approached the surface, two men, who were tried hands, were lashed with ropes and lowered over the bows, to make fast to the cable when it should appear. This was a perilous service, and the boats were there to pick up the brave fellows, if they should drop into the water. As soon as it showed itself, they dived upon it, and seizing it with their hands, fastened it with large hempen stoppers, which were quickly attached to five-inch ropes.

"It was then found, that the bight was so firmly caught in the springs of the grapnel, that one of the brave hands who put on the stoppers, was sent lower down to the grapnel, and with hammer and marlinspike, the rope was ultimately freed from the tenacious gripe of the flukes. The signal being given to haul up, the western end of the bight was cut with a saw, and grandly and majestically the cable rose up the frowning bows of the Great Eastern, slowly passing round the sheave at the bow, and then over the wheels on to the fore part of the deck. The greatest possible care had to be taken by Mr. Canning and his assistants, to secure the cable by putting on stoppers, and to watch the progress of the grapnel, rope, and shackles, round the drum, before it received the cable itself."

When once it was made fast, all took a long breath. The cable was recovered. They had the sea-serpent at last. There the monster lay, its neck firmly in their gripe, and its black head lying on the deck. But even then there was no cheering, as when they caught it two weeks before. Men are sometimes stunned by a sudden success, and hardly know if it be not all a dream. So now they looked at the cable with eager eyes, but without a word, and some crept toward it to take it in their hands, to be sure that they were not deceived. Yes—it was the same that they paid out into the sea thirteen months before!

But their anxiety was not over. Now that they had regained the lost cable of 1865, was it good for any thing? It had been lying more than a year at the bottom of the deep. What if it should prove to have been broken somewhere in the eleven hundred miles between the ship and Ireland? What if some sharp rock had worn it away, or some marine insect had eaten into its heart? If there were but a pin's point, anywhere in its covering of flesh, the vital current might escape through it into the sea. Fears like these restrained their exultation. It was yet too soon to proclaim their victory. So, as the cable was passed along the deck to the testing room, where the chief electrician was to operate upon it, to see whether it was alive or dead, it was followed by an anxious group, who stood around him as he sat down at the instrument, watching his countenance as friends watch the face of a physician, when he feels the pulse of a patient to see if the heart is still beating. The scene is thus described by Mr. Robert Dudley, the artist of the expedition, whose spirited sketches in the London Illustrated News have made known to the world many incidents of this memorable voyage:

"I made my way with others, in accordance with an invitation from Willoughby Smith, to the electricians' room. Here, after another hour's preparation, during which time the cable had been carefully passed round the drums of the picking-up machinery, and a sufficient length drawn in on board, the severed end was received. And now, in their mysterious, darkened haunt, the wizards are ready to work their spells upon the tamed lightning. Not 'unholy spells' are these, or secret; for, though the wizards' den is but of limited dimensions, they have not been averse to the presence of a few visitors. Mr. Gooch is looking on; Professor Thomson, be sure, is here, a worthy 'Wizard of the North;' Cyrus Field could no more be absent than the cable itself; I think, too, Canning, hard at work as he is forward in the ship, *must* have dropped in just for a moment; Clifford, Laws, Captain Hamilton, Deane, Dudley—all have, in their several ways, a great interest in every movement of Willoughby Smith and his brother (and able assistant) Oliver; and, when the core of the cable is stripped and the heart itself—the conducting wire—fixed in the instrument, and these two electricians bend over the galvanometer in patient watching for some message from that far-off land of home to which the great news has just been signalled, then the accustomed stillness of the test-room is deepened; the ticking of the chronometer becomes monotonous. Nearly a quarter of an hour has passed, and still no sign ! Suddenly Willoughby Smith's hat is off, and the British hurrah bursts from his lips, echoed by all on board with a volley of cheers, evidently none the worse for having been 'bottled up' during the last three hours. Along the deck outside, over the ship, throughout the ship, the pent-up enthusiasm overflowed; and even before the test-room was cleared, the roaring bravos of our guns drowned the huzzas of the crew, and the whiz of rockets was heard rushing high into the clear morning sky to greet our consort-ships with the glad intelligence."

While this scene is going on on board ship, we may turn to the other end of the line. It may be well supposed that the result of this attempt was watched with deep interest at Valentia. How they looked for the first signal from the deep, and how the tidings came, is thus told in the London Spectator:

"Night and day, for a whole year, an electrician has always been on duty, watching the tiny ray of light through which signals are given, and twice every day the whole length of wire—one thousand two hundred and forty miles—has been tested for conductivity and insulation. . . . The object of observing the ray of light was of course not any expectation of a message, but simply to keep an accurate record of the condition of the wire. Sometimes, indeed, wild, incoherent messages from the deep did come, but these were merely the results of magnetic storms and earth-currents, which deflected the galvanometer rapidly, and *spelt the most extraordinary words, and sometimes even sentences of nonsense.* Suddenly, last Sunday morning, at a quarter to six o'clock, while the light was being watched by Mr. May,* he observed a peculiar indication about it, which showed at once to his experienced eye that a message was at hand. In a few minutes afterward the unsteady flickering was changed to coherency, if we may use such a term, and at once the cable began to speak, to transmit, that is, at regular intervals, the appointed signals which indicated human purpose and method at the other end, instead of the hurried signs, broken speech, and inarticulate cries of the illiterate Atlantic. After the long interval in which it had brought us nothing but the moody and often delirious mutterings of the sea, stammering over its alphabet in vain, the words 'Canning to Glass' must have seemed like the first rational word uttered by a high-fevered patient, when the ravings have ceased and his consciousness returns."

The telegraphic fleet remained together but a few hours after this recovery of the lost cable. The battle was gained, and the three ships were no longer needed. The Albany, therefore, parted company to pick up the buoys, and at once sailed for England, while the Great Eastern, attended by the faithful Medway, turned to the west. It was about nine o'clock that the ship

* This is an error. Mr. Crocker, an operator in the Telegraph House at Valentia, was the fortunate one on watch at that hour, on whose eye the first ray fell, as a spark of life from the dead.

RECOVERY OF THE LOST CABLE.

began to pay out the cable. Up to that time it had continued calm, but the morning was raw and chill, and the sea began to rise as if in anger at those who had torn from it its prey. Captain Anderson looked anxiously at the signs of the coming storm. It seemed as if Heaven had kept back the winds during the critical day and night when they were lifting the cable! But now the tempest was upon them, and for thirty-six hours it swept the ocean. All trembled lest they should not be able to hold on. But little incidents sometimes turn the current of one's thoughts, and give a feeling of peace even in the midst of anxiety. Says Mr. Field:

"In the very height and fury of the gale, as I sat in the electrician's room, a flash of light came up from the deep, which having crossed to Ireland, came back to me in mid-ocean, telling that those so dear to me, whom I had left on the banks of the Hudson, were well, and following us with their wishes and their prayers. This was like a whisper of God from the sea, bidding me keep heart and hope. The Great Eastern bore herself proudly through the storm, as if she knew that the vital cord which was to join two hemispheres, hung at her stern; and so on Saturday, the seventh of September, we brought our second cable safely to the shore."

The scene at Heart's Content, when the fleet appeared the second time, was one that beggars description. Its arrival was not unexpected, for the success on Sunday morning, that had been telegraphed to Ireland, was at once flashed across the Atlantic, and the people were watching for its coming. As the ships came up the harbor it was covered with boats, and all were wild with excitement; and when the big shore-end was got out of the Medway, and dragged to land, the sailors hugged it and almost kissed it in their extravagance of joy; and no sooner was it safely landed than they seized Mr. Field, Mr. Canning, and Mr. Clifford in their arms, and raised them over their heads, while the crowd cheered with tumultuous enthusiasm.

The voyage of the Great Eastern was ended. Twice had she been victorious over the sea; twice had she laid the spoils of victory on the shores of the New World, and her mission was accomplished. All on board, who had been detained weeks beyond the expected time, were impatient to return; and accordingly she prepared to sail the very next day on her homeward voyage. The Medway, which had on board the cable for the Gulf of St. Lawrence, remained two or three weeks longer, and with the Terrible, whose gallant officers had volunteered for the service, successfully accomplished that work. But the Great Eastern was bound for England, and Mr. Field had now to part from his friends on board. It was a trying moment. Rejoiced as he was at the successful termination of the voyage, yet when he came to leave the ship, where he had spent so many anxious days and weeks, both this year and the year before; and to part from men to whom he was bound by the strong ties that unite those embarked in a common enterprise—brave companions in arms—he could not repress a feeling of sadness. It was with deep emotion that Captain Anderson took him by the hand, as he said, "The time is come that we must part." As he went over the side of the ship, the commander cried, " Give him three cheers!" "And now three more for his family!" The ringing hurrahs of that gallant crew were the last sounds he heard as he sunk back in the boat that took him to the Medway, while the wheels of the Great Eastern began to move, and the noble ship, with her noble company, bore away for England.

Our story is told. We have followed the history of the Atlantic Telegraph from the beginning to the end; from the hour that the idea first occurred to its projector, turning over the globe in his library, till the cable was stretched from continent to continent. Between these two points of time many years have passed, and many struggles intervened. Never did an enterprise pass through more vicissitudes; never was courage tried by more reverses and disappointments, the constant repetition of which gives to this narrative an almost painful interest. Yet that background of disaster only sets in brighter relief the spirit that bore up under all, the faith that never despaired, and the patience that was never weary. It was a pathetic as well as heroic story which Mr. Field had to tell when it was all over. He said:

"It has been a long, hard struggle. Nearly thirteen years of anxious watching and ceaseless toil. Often my heart has been ready to sink. Many times, when wandering in the forests of Newfoundland, in the pelting rain, or on the deck of ships, on dark, stormy nights—alone, far from home—I have almost accused myself of madness and folly to sacrifice the peace of my family, and all the hopes of life, for what might prove after all but a dream. I have seen my companions one and another falling by my side, and feared that I too might not live to see the end. And yet one hope has led me on, and I have prayed that I might not taste of death till this work was accomplished. That prayer is answered; and now, beyond all acknowledgments to men, is the feeling of gratitude to Almighty God." *

"Long and hard" indeed had been the way, but in the end what a triumph was gained: an achievement that was one of the most marvellous in all history, as a proof of man's dominion over the forces of nature. When it was first proposed to span the Atlantic, it seemed but a beautiful dream, fascinating indeed to the imagination, but beyond all human power: and men listened to the picture of what might be with delighted amazement and wondering incredulity. In an oration at the opening of the Dudley Observatory at Albany, in 1857, Edward Everett spoke thus of the projected Atlantic Telegraph:

* Speech at the Chamber of Commerce Dinner, Nov. 15, 1866.

"I hold in my hand a portion of the identical electrical cable, given me by my friend Mr. Peabody, which is now in progress of manufacture to connect America with Europe. Does it seem all but incredible to you that intelligence should travel for two thousand miles, along those slender copper wires, far down in the all but fathomless Atlantic, never before penetrated by aught pertaining to humanity, save when some foundering vessel has plunged with her hapless company to the eternal silence and darkness of the abyss? Does it seem, I say, all but a miracle of art, that the thoughts of living men—the thoughts that we think up here on the earth's surface, in the cheerful light of day—about the markets and the exchanges, and the seasons, and the elections, and the treaties, and the wars, and all the fond nothings of daily life, should clothe themselves with elemental sparks, and shoot with fiery speed, in a moment, in the twinkling of an eye, from hemisphere to hemisphere, far down among the uncouth monsters that wallow in the nether seas, along the wreck-paved floor, through the oozy dungeons of the rayless deep; that the latest intelligence of the crops, whose dancing tassels will, in a few months, be coquetting with the west wind on those boundless prairies, should go flashing along the slimy decks of old sunken galleons, which have been rotting for ages; that messages of friendship and love, from warm, living bosoms, should burn over the cold, green bones of men and women, whose hearts, once as warm as ours, burst as the eternal gulfs closed and roared over them centuries ago!"

But a few years passed, and the vision became a reality. The heart of the world beat under the sea.

CHAPTER XVIII.

THE AFTERGLOW.

It is the clear shining after rain. The storms that swept the sea, have blown themselves out, and all is tranquil on the face of the deep. The cable is lying in its ocean bed uniting the two hemispheres, nevermore to be separated. And now comes the public recognition on both sides the Atlantic, though in different form. The event had produced a profound impression throughout the civilized world. Yet it was a singular illustration of the changes in public interest, that, whereas in 1858 a temporary success had kindled the wildest enthusiasm in the United States, while in England it was regarded almost with indifference, now the state of feeling in the two countries was completely reversed. In Great Britain it was the theme of boundless congratulation, while in America the public mind—dulled perhaps by the excitements of four years of war—received the news with composure. The reason was, in part, that England had had a larger share in the later than in the earlier expeditions. Certainly none could deny the inestimable services rendered by her men of science, her seamen, her engineers, and her great capitalists; and it was most fit that the country which they had honored should do them honor in its turn. Scarcely had the Great Eastern recrossed the sea before those to whom the empire owed so much, were duly recognized in the following letter from the Earl of Derby, then Prime Minister, addressed to Sir Stafford Northcote, who was to preside at a dinner given in Liverpool, to celebrate the great achievement:

"BALMORAL, Saturday, Sept. 29, 1866.

"DEAR SIR STAFFORD : As I understand you are to have the honor of taking the chair at the entertainment which is to be given on Monday next, in Liverpool, to celebrate the double success which has attended the great undertaking of laying the cable of 1866, and recovering that of 1865, by which the two continents of Europe and America are happily connected, I am commanded by the Queen to make known to you, and through you to those over whom you are to preside, the deep interest with which Her Majesty has regarded the progress of this noble work ; and to tender Her Majesty's cordial congratulations to all of those whose energy and perseverance, whose skill and science have triumphed over all difficulties, and accomplished a success alike honorable to themselves and to their country, and beneficial to the world at large. Her Majesty, desirous of testifying her sense of the various merits which have been displayed in this great enterprise, has commanded me to submit to her, for special marks of her royal favor, the names of those who, having had assigned to them prominent positions, may be considered as representing the different departments, whose united labors have contributed to the final result; and Her Majesty has accordingly been pleased to direct that the honor of knighthood should be conferred upon Captain Anderson, the able and zealous commander of the Great Eastern; Professor Thomson, whose distinguished science has been brought to bear with eminent success upon the improvement of submarine telegraphy; and on Messrs. Glass and Canning, the manager and engineer respectively of the Telegraph Maintenance Company, whose skill and experience have mainly contributed to the admirable construction and successful laying of the cable. Her Majesty is further pleased to mark her approval of the public spirit and energy of the two companies who have had successively the conduct of the undertaking, by offering the dignity of a baronetcy of the United Kingdom to Mr. Lampson, the Deputy Chairman of the original company, to whose resolute support of the project in spite of all discouragements it was in a great measure owing that it was not at one time abandoned in despair; and to Mr. Gooch, M.P., the Chairman of the company which has finally completed the design. If among the names thus submitted to and approved by Her Majesty, that of Mr. Cyrus Field does not appear, the omission must not be attributed to any disregard of the eminent services which, from the first, he has rendered to the cause of transatlantic telegraphy, and the zeal and resolution with which he has adhered to the prosecution of his object, but to an apprehension lest it might appear to encroach on the province of his own Government, if Her Majesty were advised to offer a citizen of the United States, for a service rendered alike to both countries, British marks of honor, which, following the example of another highly distinguished citizen, he might feel himself unable to accept."

The reason assigned by Lord Derby for the omission of Mr. Field's name in the distribution of honors, was perfectly understood and entirely satisfactory. The British Government had once before offered a baronetcy to Mr. George Peabody in recognition of his princely benefactions to the poor of London, but while he appreciated the honor, he felt that as a citizen of the United States, he could not accept it, and the same reason would apply in the present case. But while this alone prevented official recognition, it could not prevent the hearty expression of Englishmen who knew the history of the great enterprise from the beginning. At this very dinner, the Chairman gave, as the first toast, "The Original Projectors of the Atlantic Cable," which he proposed early in order to give Mr. Cyrus Field (who was very near to them, although he happened to be in America!) a chance of responding! The allusion is explained by the remark of one present who had said : —

"You will be pleased to hear that Mr. Bright has kindly brought the telegraph wire into the room in which we are sitting, and no sooner will the toast involving the mention of Mr. Field's name be given from the chair, than it will be flashed with lightning speed to Valentia, thence to Newfoundland, and if Mr. Field is at home, it is quite possible that he himself will receive it, ere the echo of your ringing cheers has died away in Liverpool."

A message was at once sent from the room to Newfoundland, and a reply received back that Mr.

Field had left for New York. In continuing his speech, Sir Stafford Northcote said: "I think there can be no doubt in the minds of those who have carefully examined the history of these transactions, that it is to Mr. Cyrus Field that we owe the practical carrying out of the idea which has borne such glorious fruit. I am sure there is none to whom we owe more, or whose name stands in prouder connection with this great undertaking, than the name of Mr. Cyrus Field."

He called upon Sir Charles Bright to reply, who detailed somewhat the history of the enterprise from the very beginning in 1856, when "Mr. Cyrus Field, to whom the world was more indebted than to any other person for the establishment of the line, came to England upon the completion of the telegraph between Nova Scotia and Newfoundland."

To the same effect is the testimony of a distinguished writer, W. H. Russell, LL.D., who was on board of the Great Eastern in 1865, as the correspondent of The Times, and wrote a very graphic History of the Expedition (p. 10):

"It has been said that the greatest boons conferred on mankind, have been due to men of one idea. If the laying of the Atlantic cable be among those benefits, its consummation may certainly be attributed to the man who, having many ideas, devoted himself to work out one idea, with a gentle force and patient vigor which converted opposition and overcame indifference. Mr. Field may be likened either to the core, or the external protection, of the cable itself. At times he has been its active life ; again he has been its iron-bound guardian. Let who will claim the merit of having first said the Atlantic cable was possible ; to Mr. Field is due the inalienable merit of having made it possible, and of giving to an abortive conception all the attributes of healthy existence."

Sir William Thomson, on the final triumph, wrote :

"My dear Field, I cannot refrain from putting down in black and white my hearty congratulations on your great success. Few know better than I do how well you deserve it."

Eight months after he wrote from Scotland :

"I am sorry I had not an opportunity of saying in public how much I value your energy and perseverance in carrying through the great enterprise, and how clearly you stand out in its history as its originator and its mainspring from beginning to end."

Next to Sir William Thomson was Mr. C. F. Varley, who was associated in the work from an early day, and did much to solve the difficult problems of ocean telegraphy, and who wrote to Mr. Field, speaking from his personal knowledge : "You did more than any other to float the concern, and single-handed saved the whole scheme from collapse more than once."

Captain Sir James Anderson repeated the same conviction in numberless forms. He had seen how the presence of Mr. Field in London instantly revived the languid enthusiasm of others, and infused his own energy into the enterprise, and declared again and again that but for these heroic and incessant efforts the whole scheme would have broken down, and been delayed for many years.

Such expressions from English associates in the great work might be multiplied to any extent. They are much stronger than any language used by the author of this volume, who has purposely kept back such testimonies, lest it should seem that he wished to exalt an individual, when he sought to do justice to all, on both sides of the Atlantic.

Nor was such recognition confined to England. The King of Italy conferred on Mr. Field the cross of the order of St. Mauritius, as an acknowledgment from the country of Columbus to one who had done so much to unite to the Old World that New World which Columbus discovered.

A still higher honor was paid by the Great Exposition in Paris, in 1867, which, gathering the products of the genius and skill and industry of all nations, recognized the labors of men of all countries, who, by their discoveries or great enterprises, had rendered eminent services to the cause of civilization. It awarded the GRAND PRIZE, the highest distinction it had to bestow, to Mr. Field by name, jointly with the Anglo-American and Atlantic Telegraph Companies, thus recognizing, as was most due, the splendid exhibition of the science and the capital of England, which were

THE AFTERGLOW. 381

never more directly employed for the benefit of the human race, than in the uniting of the two Hemipheres, while it gave the first place in the grand design to its American leader.

But to an American no praise is so dear as that which comes from his own countrymen. First of all to Mr. Field, was that which came from the faithful few who had stood by him and witnessed his exertions for twelve long years. At the first annual meeting of the stockholders of the New York, Newfoundland, and London Telegraph Company, the following resolution was, on motion of Mr. Moses Taylor, seconded by Mr. Wilson G. Hunt, unanimously adopted:

Whereas, This Company was the first ever formed for the establishment of an Atlantic Telegraph; an enterprise upon which it started in the beginning of 1854, at the instance of Mr. Cyrus W. Field, and which, through his wise and unwearied energy, acting upon this Company, and others afterwards formed in connection with it, has been successfully accomplished: Therefore the stockholders of this Company, at this their first meeting since the completion of the enterprise, desiring to testify their sense of Mr. Field's services:

Resolve: First—That to him more than any other man, the world is indebted for this magnificent instrument of good; and but for him it would not, in all probability, be now in existence;

Second—That the thanks of the stockholders of this Company are hereby given to Mr. Field for these services, which, though so great in themselves, and so valuable to this Company, were rendered without any remuneration; and

Third—That a copy of this resolution, certified by the Chairman and Secretary of this meeting, be delivered to Mr. Field as a recognition, by those who best know, of his just right to be always regarded as the first projector, and most persistent and efficient promoter, of the Atlantic Telegraph.

PETER COOPER, *Chairman.*
WILSON G. HUNT, *Secretary.*

To testify the public appreciation of this great achievement, and of his part in it, the Chamber of Commerce of New York invited Mr. Field to a public banquet, which was given on the fifteenth of November. It was attended by about three hundred gentlemen—not only merchants and bankers, but men of all professions—lawyers and judges, clergymen and presidents of colleges, members of the Government and foreign ministers, and officers of the army and navy. The President of the Chamber of Commerce, Mr. A. A. Low, presided, and, at the close of his opening speech, said:

" We may fairly claim that, from first to last, Cyrus W. Field has been more closely identified with the Atlantic Telegraph than any other living man; and his name and his fame, which the Queen of Great Britain has justly left to the care of the American government and people, will be proudly cherished and gratefully honored. We are in daily use of the fruits of his labors; and it is meet that the men of commerce, of literature and of law, of science and art—of all the professions that impart dignity and worth to our nature—should come together and give a hearty, joyous, and generous welcome to this truly chivalrous son of America."

He proposed the health of their guest:

"CYRUS W. FIELD, the projector and mainspring of the Atlantic Telegraph: while the British government justly honors those who have taken part with him in this great work of the age, *his* fame belongs to us, and will be cherished and guarded by his countrymen."

In his reply, Mr. Field told the story with the utmost simplicity, passing rapidly over the nearly thirteen years, through which the enterprise had struggled with such doubtful fortunes, and taking pains to do full justice to all who shared in its labors, its disappointments and its triumphs. Especially did he award the highest praise to the government of England for its liberal and constant support; to her men of science and her great capitalists, and to the officers of ships, electricians and engineers, who had taken part in this undertaking. In closing, he said:

"Of the results of this enterprise—commercially and politically—it is for others to speak. To one effect only do I refer as the wish of my heart—that, as it brings us into closer relations with England, it may produce a better understanding between the two countries. Let who will speak against England—words of censure must come from other lips than mine. I have received too much kindness from Englishmen to join in this language. I have eaten of their bread and drunk of their cup, and I have received from them, in the darkest hours of this enterprise, words of cheer which I shall never forget; and if any words of mine can tend to peace and good will, they shall not be wanting. I beg my countrymen to remember the ties of kindred. Blood is thicker than water. America with all her greatness has come out of the loins of England; and though there have been sometimes family quarrels—bitter as family quarrels are apt to be—still in our hearts there is a yearning for the old home, the land of our fathers; and he is an enemy of his country and of the human race, who would stir up strife between two nations that are one in race, in language and in religion. I close with this sentiment : ENGLAND AND AMERICA—CLASPING HANDS ACROSS THE SEA, MAY THIS FIRM GRASP BE A PLEDGE OF FRIENDSHIP TO ALL GENERATIONS!" (To which the whole assembly responded by rising, and by prolonged and tumultuous cheers.)

In the brilliant array of guests was recognized the tall form of General Meade, who was loudly called for as " the hero of Gettysburg," to which he replied that there was but one hero on this occasion, and he had travelled a hundred miles to be there that night to do him honor. He said : " I have watched with eagerness the struggle through which he has passed and the disasters which attended his early efforts ; and I have admired and applauded, from the bottom of my heart, the tenacity of purpose with which that man has continued to hold on to his original idea, with a firm faith to carry to completion one of the greatest works the world has ever seen."

The heartiness of this soldierly reply was echoed by the bluff old warrior, Admiral Farragut, who had been so often through the smoke and flame of battle, that he knew how to appreciate not only common courage, but the desperate tenacity which holds on in spite of disaster, that has gained many a victory.

Letters were read from the President of the United States, from Chief Justice Chase, from General Grant, from Sir Frederick Bruce, the British Minister, from Senators Morgan and Sumner, from General Dix, Minister to France, and others. The Chief Justice of the United States wrote:

"I am very sorry that I cannot leave Washington this week, and so cannot avail myself of your kind invitation to join you in congratulations to Mr. Field upon the success of his grand undertaking. It is the most wonderful achievement of civilization; and to his sagacity, patience, perseverance, courage, and faith, is civilization indebted for it.

"Such works entitle their authors to distinguished rank among public benefactors. You will write the name of your honored guest high upon that illustrious roll, and there it will remain in honor, while oceans divide and telegraphs unite mankind."

There was a telegraph instrument in the room, and despatches were received during the evening from Mr. Seward, Secretary of State, and other members of the Cabinet at Washington, from Lord Monck, Governor-General of Canada, from the Governor of Newfoundland, and others. One, from Captain Sir James

Anderson, was dated at London the same day. John Bright also wrote a despatch and sent it to London, but by an oversight it was not forwarded. He afterward wrote a letter, giving the message. It was as follows:

"It is fitting you should honor the man to whom the whole world is debtor. He brought capital and science together to do his bidding, and Europe and America are forever united. I cannot sit at your table, but I can join in doing honor to Cyrus W. Field. My hearty thanks to him may mingle with yours."

He adds that he regarded what had been done as the most marvellous thing in human history; as more marvellous than the invention of the art of printing, or, he was almost ready to say, than the voyages of the Genoese; and of Mr. Field, he says, " The world does not yet know what it owes to him, and this generation will never know it."

About the same time, in a speech at a great Reform Meeting in Leeds, he bore this proud testimony:

" A friend of mine, Cyrus Field, of New York, is the Columbus of our time, for after no less than forty voyages across the Atlantic, in pursuit of the great aim of his life, he has at length, by his cable, moored the New World close alongside the Old."

Nor was this mere rhetoric, a burst of extravagance, to which an orator might give way in the excitement of a public occasion; it was a comparison which he repeated on many occasions, though slightly varied in expression. Mr. G. W. Smalley, the well-known correspondent of the New York Tribune, in writing from London, on the very day that Mr. Field was carried to his grave, recalls how he heard it from Mr. Bright's own lips. He says:

"The great orator spoke of the great American in terms which he did not bestow lavishly, and never bestowed carelessly. His respect for Mr. Field's public work was sufficiently shown in the splendid eulogy which he passed upon him. To be called by such a man as Mr. Bright the Columbus of the Nineteenth Century is renown enough for any man. The epithet is imperishable. It is, as Thackeray said of a similar tribute to Fielding in Gibbon, like having your name written on the dome of St. Peter's. The world knows it and the world remembers. I heard Mr. Bright use the phrase, and he adorned and emphasized it in his noblest tones."

America has no official honors to bestow, no knighthoods or baronetcies to confer. But one honor it has, the thanks of Congress, which, like the thanks of Parliament, is the more highly prized in that it is so rarely bestowed, being reserved generally for distinguished officers in the army or navy, like Generals Grant, Sherman or Sheridan, or Admiral Farragut, who have won great victories. Yet such was the feeling on this occasion, that when Senator Morgan, of New York, moved a vote of thanks in the name of the country, it met with an immediate response. It was at once referred to the Committee on Foreign Relations, which reported unanimously in its favor; and when, some weeks after, giving time for due deliberation, it was brought up for action, it passed with entire unanimity. In the House of Representatives it was preceded by many bills, so that there was danger that it might not be reached before the end of the session, yet on the very last day Speaker Colfax requested unanimous consent of the House to take it up out of its order, which was granted, and the resolution was then read three times, and passed unanimously. It is as follows:

"*Resolved, by the Senate and House of Representatives of the United States of America, in Congress assembled,* That the thanks of Congress be, and they hereby are, presented to Cyrus W. Field of New York, for his foresight, courage, and determination in establishing telegraphic communication by means of the Atlantic cable, traversing mid-ocean and connecting the Old World with the New ; and that the President of the United States be requested to cause a gold medal to be struck, with suitable emblems, devices, and inscriptions, to be presented to Mr. Field.

"*And be it further resolved,* That when the medal shall have been struck, the President shall cause a copy of this joint resolution to be engrossed on parchment, and shall transmit the same, together with the medal, to Mr. Field, to be presented to him in the name of the people of the United States of America.

"Approved March 2, 1867.

"ANDREW JOHNSON."

This action of Congress reached Mr. Field in England. As he was about returning to America, Lord Derby, still at the head of the government, addressed to him a letter in which he repeated what he had said before "in the Queen's name," "how much of the success of the great undertaking of laying the Atlantic Cable was due to the energy and perseverance with which, from the very first, in spite of all discouragements, you adhered to and supported the project;" and adding, "Your signal services in carrying out this great undertaking have been already fully recognized by Congress; and it would have been very satisfactory to the Queen to have included your name among those on whom, in commemoration of this great event, her Majesty was pleased to bestow British honors, if it had not been felt that, as a citizen of the United States, it would hardly have been competent to you to accept them. As long, however, as the telegraphic communication between the two Continents lasts, your name cannot fail to be honorably associated with it."

This surely was all that could be expected from the government, but some there were in England who felt that there was still a debt of honor to be paid, which required some public testimonial. Accordingly, on Mr. Field's return to London, in 1868, they prepared for him an imposing demonstration in the form of a banquet, given at Willis's Rooms, on the first of

July, at which was assembled one of the most distinguished companies that ever met to do honor to a private citizen of any country. It embraced over four hundred gentlemen of all ranks: ministers of state, members of parliament, both Lords and Commons; officers of the army and navy; great capitalists—merchants and bankers; men of science and of letters; inventors, electricians, and engineers—men eminent in every walk of life. The Duke of Argyll presided, and speeches were made by three members of the government—Sir John Pakington, Secretary of State for War; Sir Stafford Northcote, Secretary of State for India; and Sir Alexander Milne, First Sea Lord of the Admiralty; by John Bright; by the venerable Lord Stratford de Redcliffe, so long the British Minister at Constantinople; and by M. de Lesseps, the projector of the Suez Canal, who had come from Egypt expressly to be present. It was a tribute such as is rarely paid to any man while living—such tributes being reserved for the dead—and is still more honorable in this case, alike to the givers and the receiver, in that it was paid by the people of one country to a citizen of another, who was regarded in both as their common benefactor.

But enough of praise that can fall only on the dull, cold ear of death. A few words on the after years of this busy life, and I have done. These years brought a rich reward for all the sacrifices of the past. The first feeling was one of infinite relief that at last the victory was won. The terrible strain was taken off, and to him who had borne it so long, the change to the quiet of his own happy home was inexpressibly grateful after his many and long separations. He was now in his own country and under his own roof, but with a name that was known on both sides of the sea. The complete success of the Atlantic Telegraph had given him an immense reputation at home and abroad. It seemed as if the struggles of life were all over, leaving only its honors to be enjoyed. What more could he ask to make life worth living than the respect of his countrymen for his courage, energy and perseverance, and a name honored all over the civilized world as one of the world's benefactors?

The practical results of the cable were even greater than he had dared to anticipate. In the space of a few months it wrought a commercial revolution in America. It was a new sensation to have the Old World brought so near, that it entered into one's daily life. Every morning, as Mr. Field went to his office, he found laid on his desk at nine o'clock the quotations on the Royal Exchange at twelve! Lombard Street and Wall Street talked with each other as two neighbors across the way. This soon made an end of the tribe of speculators who calculated on the fact that nobody knew at a particular moment the state of the market on the other side of the sea, an universal ignorance by which they profited by getting the earliest advices. But now everybody got them as soon as they, for the news came with the rising of each day's sun, and the occupation of a class that did much to demoralize trade on both sides of the ocean was gone.

The same restoration of order was seen in the business of importations, which had been hitherto almost a matter of guess-work. A merchant who wished to buy silks in Lyons, sent out his orders months in advance, and of course somewhat at random, not knowing how the market might turn, so that when the costly fabrics arrived, he might find that he had ordered too many or too few. A China merchant sent his ship round the world for a cargo of tea, which returned after a year's absence, bringing not enough to supply the public demand, leaving him in vexation at the thought of what he might have made, "if he had known," or, what was still worse, bringing twice too much, in which case the unsold half remained on his hands. This was a risk against which he had to be insured, as much as against fire or shipwreck. And the only insurance he could have was to take reprisals by an increased charge on his unfortunate customers.

This double risk was now greatly reduced, if not entirely removed. The merchant need no longer send out orders a year beforehand, nor order a whole shipload of tea when he needed only a hundred chests, since he could telegraph to his agent for what he wanted

THE AFTERGLOW.   393

and no more. With this opportunity for getting the latest intelligence, the element of uncertainty was eliminated, and the importer no longer did business at a venture. Buying from time to time, so as to take advantage of low markets, he was able to buy cheaper, and of course to sell cheaper. It would be a curious study to trace the effect of the cable upon the prices of all foreign goods. A New York merchant, who has been himself an importer for forty years, tells me that the saving to the American people cannot be less than many millions every year.

But the slender cord beneath the sea had finer uses than to be a reporter of markets, giving quotations of prices to counting rooms and banking houses; it was a link between hearts and homes on opposite sides of the ocean, bearing messages of life and death, of joy and sorrow, of hopes and fears. One of its happiest uses was the relief of anxiety. A ship sailed for England with hundreds of passengers, but did not arrive at her destination on the appointed day. Instantly a thousand hearts were tortured with fear, lest their loved ones had gone to the bottom of the sea, when the cable reported that the delay was due simply to an accident to her machinery, that would keep her back for a day or two, but that the good ship was safe with all on board. What arithmetic can compute the value of a single message that relieves so much anguish? Thus the submarine telegraph stretched out its long arms under the sea, to lay a friendly hand on two peoples, and give assurance to both.

Such a triumph of commercial enterprise was enough to satisfy the pride and ambition of any man; but it was not in Mr. Field's nature to rest content with any success, however great, and he was always reaching out for some new undertaking to give scope to his restless activity. Such an opportunity he found in giving rapid transit to New York, a city which, though it has one of the finest harbors in the world, with approaches from the sea that afford every possible advantage for commerce, is not so favorably situated landward, as it is built on a long and narrow island, between two broad rivers, which confine it on either side, so that it is stretched out to such distances that it is no easy matter to pass from one end to the other. From the Battery to the Harlem river is ten miles, so that working men, who lived so far away, were an hour or more in getting from their homes to their place of work, and as long in getting back again, a large inroad upon their hours of rest or domestic comfort. The only means of transportation was by street cars, which moved slowly, and in winter, whenever the streets were blocked with snow, were crowded to suffocation, and dragged at a snail's pace to the upper end of the island.

This was the great barrier to the city's growth, and must be removed if it was not to be stunted and dwarfed by these limitations. To furnish some relief, an elevated railroad, built on stilts, had been attempted on a small scale, but soon broke down, when Mr. Field bought the control of the whole concern, and took it in his own strong arms. It was no easy matter to galvanize it into life, for though it had a charter, it was still obstructed in the legislature, and in the courts, so that it was a long time before he could get full possession. But once master of the situation, he undertook the work on a grand scale, and pushed it with such vigor that in less than two years the road was in operation. It has since been extended with the public demand, until now (in 1892) there are thirty-six miles of road, over which the trains sweep incessantly from the bay to the river, and from the river to the bay.

The structures are not indeed the most graceful in the world, as they bestride the long avenues of the city. But these tall iron pillars, that line our streets for miles, are the long legs of civilization, and have a somewhat imposing effect as they stretch away into the distance, with the fire-drawn cars flying swiftly over them. Dean Stanley glorified them by a historical parallel which could occur only to one full of the wonders of ancient times, that started into life under the touch of his imagination. Going with him one day on an excursion, he stepped briskly (for his frame was so light as to offer little impediment to motion), and as he mounted the long stairway, and stood on the platform above the crowded street below, he exclaimed,

"This is Babylonian! Four chariots driving abreast on the walls of the city!"

But Babylonian or American, the success was enormous. As soon as the public became familiar with these elevated roads, and felt that they were safe as well as swift, the people swarmed upon them, in numbers constantly increasing, till now they carry over seven hundred thousand passengers a day! On the day of the Columbus celebration (October 12th) it was a million and seventy-five thousand! Indeed, if we are not staggered by numbers, we may sum up the whole in the amazing statement demonstrated by figures, that since these roads were opened, they have carried over eighteen hundred millions of passengers, more than the whole population of the globe!

Nor should it be forgotten that, not only is the facility they afford the greatest, but the fares the lowest, for, thanks to Mr. Field, they were reduced years ago to five cents at all hours and for the longest distance, the ten miles from the Battery to the Harlem river.

The effect was immediate in the appreciation of real estate in the city, the assessed value of which has already advanced by the sum of five hundred millions of dollars! The increased taxation is enough to pay for all the cost, while as a relief to the congested parts of the city, and as furnishing a means for that easy circulation, which is as necessary to a great city as a free circulation of the blood is to the human body, it is not too much to say that the construction of the elevated railroads is the greatest material benefit that has ever been conferred on the city of New York.

But busy as Mr. Field was through all these years, much of his life was spent abroad. He had interests on both sides of the Atlantic, but stronger than his interests were his friendships to attract him across the sea. He had come to feel as much at home in England as in his own country: and his visits were so frequent that his sudden appearances and disappearances were a subject of amused comment to his English friends. When Dean Stanley was in America, a reception was given to him at the Century Club, where in a very happy address, he referred to the ties between the two countries, among which was "the wonderful cable, on which it is popularly believed in England, that my friend and host, Mr. Cyrus Field, passes his mysterious existence, appearing and reappearing at one and the same moment in London and in New York!"

As Mr. Field was thus brought near to his English friends, they in turn were brought near to him, for as no man in America was better known abroad, no house received more foreign guests, many of whom he had not met before, but who brought letters to him, and there was no end to his hospitality. John Bright he could not persuade to cross the sea; but he had the pleasure to welcome his co-laborer in the repeal of the Corn-laws, Richard Cobden. The house in Gramercy

Park became famous for its receptions. Many will recall that given to the Marquis of Ripon and the other High Commissioners, who came a year or two after the war, as representatives of the British government, and negotiated at Washington the treaty which settled the Alabama claims; and those to Dean Stanley and Archdeacon Farrar; and to many others. If the strangers happened to arrive in the summer time, they were entertained at his beautiful country seat on the Hudson, to which he had given the name of "Ardsley," from the seat of John Field the astronomer, who lived in the West Riding of Yorkshire more than three centuries ago, and introduced the Copernican astronomy into England, and from whom the family are descended.

In some cases when he went abroad, England was but the starting point for excursions on the Continent, in which he visited almost every European country. In 1874, in company with two well-known Americans, Bayard Taylor and Murat Halstead, he made a voyage to Iceland, as ten years before he had been to Egypt, as a delegate from the New York Chamber of Commerce, to witness the opening of the Suez Canal.

In 1880–81 he took a still longer flight around the world. Waiting till after the Presidential election, that he might cast his vote for his friend General Garfield, the very next day he left with his wife in a special car for San Francisco, where after a few days, they took ship for Japan, from which they passed through the Inland Sea to Shanghai, and from China to Singapore, and up the Bay of Bengal to Calcutta, where he found the same English nobleman whom he had entertained in New York, the Marquis of Ripon, Governor-General of India. Going up the country, the travellers visited Agra and Delhi, where the wonders of architecture showed the magnificence of the old Mogul Empire. The whole journey was one of infinite pleasure and instruction, and they were never weary of talking of the strange manners and customs of the people of Asia.

When they returned to America, General Garfield was President of the United States, who, though a Western man by birth, had been educated in New England, at Williams College in Massachusetts, where he had been graduated twenty-five years before, and which he had a desire to revisit; and it was arranged that he should leave Washington in the morning of July 2d, with as many of his cabinet as could be spared from the seat of government, and come on to New York and all be entertained at "Ardsley," and the next day proceed up the Hudson and across the country to Williamstown; a programme which was interrupted by the terrible news that on arriving at the station in Washington he had been shot, an event that instantly recalled the assassination of Lincoln. At once there rose a cry of horror from one end of the land to the other, and for weeks the whole country was watching by the bedside of the illustrious sufferer.

Of course, the sympathy for the wife and children was universal, but Mr. Field was the first to give this sympathy a practical direction. With his quick eye he saw the condition in which they would be left by the death of the President, as for them the law makes no provision. His salary stops at the very day and hour that he ceases to live, nor is there a pension settled upon his family, nor can anything be paid from the national treasury except by special act of Congress. In this extremity it occurred to Mr. Field that what the Government failed to do should be made up by private generosity; and even before General Garfield's death he started a subscription, heading it with five thousand dollars, and taking it in person to his rich friends. The self-imposed task occupied him several months, in which he raised a fund of over three hundred and sixty thousand dollars, which was put into United States four per cent. bonds, yielding an interest of over twelve thousand dollars a year, to be paid quarterly during the life of Mrs. Garfield, and then to go to her children. It was a great satisfaction to have thus provided for those who bore the name of a President of the United States, so that they should be able to live in the comfort and dignity that befitted the family of one who had occupied the most exalted station in the government.

Not content with this, Mr. Field went to Washington, and urged upon his friends in Congress, and finally succeeded in getting passed, a bill giving to the widows of all Presidents a pension of $5,000, which, it added to his gratification to know, would apply to the venerable Mrs. Polk: and that still goes, and will go during her lifetime, to the wife of General Grant, as the slight expression of a nation's gratitude.

Next to the interest he felt in his own country, his heart was in England. While he was an intense American, and perhaps, for that very reason, because he was an American, he claimed kindred with the people from whom we are not only descended, but have received such an inheritance of glory. In his own words: "America, with all her greatness, has come out of the loins of England." When he was in India he was proud of the mighty English race that from its little island governed an empire of two hundred and fifty millions on the other side of the globe. Some might have said that he inherited no small portion of its unconquerable spirit.

And not only did he admire Old England, but he loved Englishmen. He knew all that was said of English reserve and English pride, but long familiarity had taught him that underneath this cold exterior were many of the noblest qualities—courage, heroism and fidelity—so that it had become a part of his creed that an Englishman, when once you have won his confidence, will go farther and fight harder for a friend or for a cause than any other man on the face of the earth. Among such a people Mr. Field was proud to number many of his dearest friends.

A touching proof of their regard for him was given but a few months before his death. On the 2d of December, 1890, he and his wife celebrated their golden wedding. For fifty years they had travelled on the course of life together. Children and grandchildren had been born to them, so that at the close of half a century a large and happy family was gathered round those to whom they looked up with the tenderest affection.

Among the congratulations of that day was a large scroll, signed by Mr. Gladstone, the Duke of Argyll, Lord Monck, and some eighty others whose names are widely known. It was a graceful tribute from England to a son of America, who had done perhaps more than any other living man to bring the two countries and the two peoples together.

That golden wedding was the fit coronation of a life of wonderful activity, and all the kindred who met under that roof were grateful for the past, and full of hopes for the future.

But God's ways are not as our ways. Before many months the clouds began to gather. The next summer, when the family were all at their country home, sickness cast its shadow over their dwelling, which grew more grave till November 23d, when the leaves were falling from the trees before their door, the mother of this large household breathed her last. Two months later the eldest daughter, who was also the eldest child of the family, followed. These repeated blows fell heavy on the affectionate heart of the bereaved husband and father, and when to these were added other sorrows still, it seemed as if the clouds were piled one upon another till they darkened all the horizon. The winter was a gloomy one, from its loneliness and its many causes of sadness. But with the returning spring the grass grew green again, and the trees put forth their leaves, and it seemed as if the new life of nature must put life into the heart of man: and when he removed to the country, and began to drive about as of old among the familiar haunts, the beautiful scenery for a time delighted his eye, and the change of air brought a touch of the old spirit, as if perchance his strength were about to return. But it was only a momentary flush, and he soon took to his room, where, as he looked from his windows, and saw the sun going down over the hills beyond the Hudson, it could only remind him that for him the sun of life was about to set forever. Fair was the world without but desolate was the home within, since she who had made its brightness was gone; and here on the 12th of July 1892, the end came.

It was a beautiful morning, and the windows were

open, through which the soft summer air floated into the chamber of death, where his three brothers, all that were left of his father's family, with those of his own household, were round his bed, watching the dear pale face. Thus surrounded and beloved to the last, he ceased to breathe.

Two days later a large company from the country round and from the city gathered at Ardsley, and stood on the lawn and the slopes that lead up to the noble trees that shade the dwelling, as Bishop Potter read the blessed words, "I am the Resurrection and the Life, saith the Lord: he that believeth in Me, though he were dead, yet shall he live."

The next day we bore him away from his home, and from the great city where he had passed his busy life, back to the quiet valley where he was born, and laid him down in the shadow of the encircling hills.*
"Bury me there," he had said, "by the side of my beloved wife and by my father and mother." The earth closed over him, and all his struggles and his sorrows were buried in the grave.

The man is gone, but the work remains, a work that multiplies itself, for when once a leader and explorer had opened the way, others were swift to follow, so that now there are no less than ten cables stretched across the Atlantic, and every hour of day or night, "when men wake and when they sleep" (for even in

---

* The Berkshire Hills, Stockbridge, Massachusetts.

the hours of silence the heart is still beating, only a little more slowly), the pulse of life is kept moving to and fro. The morning news comes after a night's repose, and we are wakened gently to the new day that has dawned upon the world. That which serves to such an end; which is a connecting link between countries and races of men; is not a mere material thing, an iron chain, lying cold and dead in the icy depths of the Atlantic. It is a living, fleshly bond between severed portions of the human family, thrilling with life, along which every human impulse runs swift as the current in human veins, and will run for ever. Free intercourse between nations, as between individuals, leads to mutual kindly offices, that make those who at once give and receive, feel that they are not only neighbors but friends. Hence the "mission" of submarine telegraphy is to be the minister of peace. The first message across the deep was "Glory to God in the highest; peace on earth, good will to men," and the first news it brought was that of peace in China. And when again the sea had found a tongue, its first glad intelligence was that the great war between Austria and Prussia was ended. Thus at its very birth was this new messenger baptized in the name of Peace, and consecrated to a service worthy of its name.

> "Man marks the earth with ruin: his control
>   Stops with the shore: upon the watery plain
>   The wrecks are all thy deed."

Not all! The wrath of man adds to the fury of the elements. To strew the sea with wrecks is the work of lightning and tempest: man's nobler office is to restore what nature may destroy.

It was the chief desire of him who has gone to the grave, that the link which unites England and America might bind the countries that he loved the most in indissoluble union. Though the two nations dwell apart, on opposite shores of the same great and wide sea, they are now brought almost within the sound of each other's voice and the touch of each other's hand: they can look into each other's eyes, and exchange their morning and evening congratulations with the rising and setting of each day's sun. May the instrument through which they look and speak never startle them with rude alarms, but continue to whisper peace in tones as gentle as the murmur of the sea, as long as the winds blow and the waters roll.

APPENDIX.

INSTRUMENTS FOR SIGNALLING ACROSS THE ATLANTIC OCEAN.

IF the project of an Atlantic Telegraph be justly ascribed to the daring of an American, and its success to his courage and perseverance through years of struggle and disappointment; the solution of the scientific problem involved in it, is due to the genius of a Scotchman, whom the writer of this volume first knew (and it is a pleasant memory to have known such a man in the beginning of his splendid career) as Professor Thomson of the University of Glasgow, where his father had been professor before him, whom the son succeeded in the Department of Physics, which included the then little known science of Electricity, to which the young professor devoted himself with all the eagerness of scientific genius. The project of a telegraph across the ocean suggested new problems and new difficulties, to which he applied himself with characteristic ardor, the result of which is here given. When the second expedition of the Great Eastern (in 1866) was successful, the British Government at once recognized his eminent services; and the name of Sir William Thomson has since been recognized, among the leaders in scientific discovery, not only in England but all over the scientific world. The government has recently added a further dignity in making him a peer of the realm, an honor hitherto reserved generally for the leaders of armies, like Wellington. To confer it on a simple professor shows an advance of civilization in the respect paid to intellectual greatness. In conferring such a title, the government does not honor the man more than it honors itself. It is to the glory of England that such an honor should be paid to science in the person of Lord Kelvin, as was paid to literature in the person of Lord Tennyson.

The following, taken in substance from an English scientific review, will indicate briefly, but with sufficient clearness, the problem to be solved in signalling to great distances under the sea, and the instruments by which this is accomplished: —

The speed of signalling through a submarine cable depends upon its electrostatic capacity, which, unless it be very small, gives rise to "retardation."

In the Proceedings of the Royal Society for 1855, Sir William Thomson showed how the effect at the distant end of a cable, caused by the application of a battery at one end, could be calculated and represented graphically in what is called the "curve of arrival." After contact is first made at the sending end between the cable and one pole of the battery (the other pole being to earth), a certain interval of time elapses before any effect is felt at the distant end. This interval of time is denoted by the letter a. After the interval of time a has passed, a current begins to issue from the cable at the receiving end, and increases in strength very rapidly. After a further interval of $4a$ or after a period of $5a$ from the first application of the battery, it attains about half its maximum strength, and there is very little sensible increase in strength after a time equal to $10a$ has elapsed. The curve of arrival is drawn by taking distances along O X to represent intervals of time, and distances along O Y to represent strengths of current. Curve No. I. shows the gradual increase in strength of the received current at one end of a cable when the battery is applied to and kept in contact with the other end. For a distance corresponding to the interval of time a, the curve does not sensibly deviate from

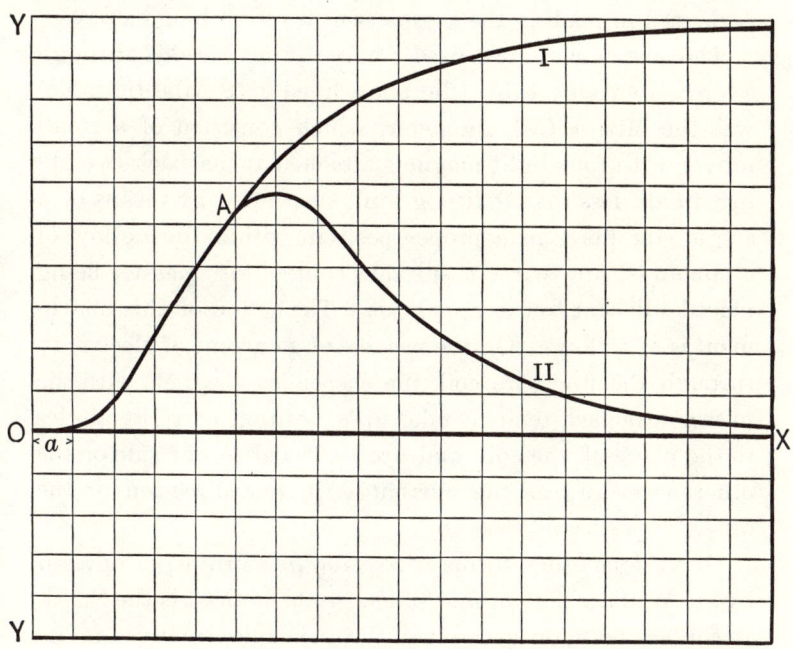

the straight line O X; in other words, no effect is observable at the receiving end during this time.

If now, instead of being continuously applied to the battery at the sending end, the cable had been applied to it during a short interval of time, and then disconnected from the battery and connected to earth, the curve of arrival would be of the form shown by curve No. II. Curve No. II. shows the effect of applying the battery during a length of time equal to 4a, and then putting the cable to earth. It will be seen that a current gradually diminishing in strength continues to flow out of the cable at the distant end for a considerable time after the battery has been disconnected. This continued discharge is what gives rise to the difficulty experienced in reading the signals sent through long cables.

The instrument first used for receiving signals through a long submarine cable (the short-lived 1858 Atlantic cable) was the Mirror Galvanometer, which consisted of a small mirror with four light magnets attached to its back (weighing, in all, less than half-a-grain), suspended by means of a single silk fibre, in a proper position within the hollow of a bobbin of fine wire: a suitable controlling magnet being placed adjacent to the apparatus. The action of this instrument is as follows. On the passage of a current of electricity through the fine wire coil, the suspended magnets with the mitror attached, tend to take up a position at right angles to the plane of the coil, and are deflected to one side or the other according as the current is in one direction or the other.

Of various other forms of *receiving* instruments devised by Sir William Thomson, the next to be noticed is the Spark Recorder, both on account of the principles involved in its construction, and because it in some respects foreshadowed the more perfect instrument, the Siphon Recorder, which he introduced some years later. The action of the Spark Recorder was as follows. An indicator, suitably supported, was caused to take a to-and-fro motion, by means of the electro-magnetic action due to the electric currents constituting the signals. This indicator was connected to a Ruhmkorff coil or other equivalent apparatus, designed to cause a

SIGNALLING UNDER THE SEA. 411

continual succession of sparks to pass between the indicator, and a metal plate situated beneath it and having a plane surface parallel to its line of motion. Over the surface of this

FIG. 1.

plate and between it and the indicator, there was passed, at a regularly uniform speed in a direction perpendicular to the line of motion of the indicator, a material capable of being acted on physically by the sparks, either through their chemical action, their heat, or their perforating force. The record of the signals given by this instrument was an undulating line of fine perforations or spots, and the character and succession of the undulations were used to interpret the signals desired to be sent.

The latest form of *receiving* instrument for long submarine cables, is that of the Siphon Recorder, for which Sir William Thomson obtained his first patent in 1867. Within the three succeeding years he effected great improvements on it, and the instrument has, since that date, been exclusively employed in working most of the more important submarine cables of the world—indeed all except those on which the Mirror-Galvanometer method is still in use.

In the Siphon Recorder (a view of which is shown in Fig. 1), the indicator consists of a light rectangular signal-coil of fine wire, suspended between the poles of a powerful electro-magnet, so as to be free to move about its longer axis which is vertical, and so joined up that the electric currents constituting the signals through the cable, pass through it. A fine glass siphon-tube is suitably suspended, so as to have only one degree of freedom to move, and is connected to the signal-coil so as to move with it. The short leg of the siphon-tube dips into an insulated ink-bottle, which permits of the ink contained by it being electrified, while the long leg is situated so that its open end is at a very small distance from a brass table, placed with its surface parallel to the plane in which the mouth of this leg moves, and over which a slip of paper may be passed at a uniform rate as in the Spark Recorder. The effect of electrifying the ink is to cause it to be projected in very minute drops from the open end of the siphon-tube, towards the brass table or on the paper-slip passing over it. Thus when the signal-coil moves in obedience to the electric signal currents passed through it, the motion then communicated to the siphon, is recorded on the moving slip of paper by a wavy line of ink marks very

FIG. 2.

close together. The interpretation of the signals is according to the Morse code; the dot and dash being represented by deflections of the line to one side or the other of the centre line of the paper.

Perfect as this instrument seemed, yet after further years of study and experiment, Sir William Thomson was able, at the close of 1883, to present to the world the Siphon Recorder, greatly improved, because in a very much simpler form. In this form of the instrument, instead of the electro-magnets,

FIG. 3.

he used two bundles of long bar-magnets of square section and made up of square bars of glass-hard steel. The two bundles are supported vertically on a cast-iron socket, and on the upper end of each is fitted a soft iron shoe, so shaped as to concentrate the lines of force and thus produce a strong

SIGNALLING UNDER THE SEA.

magnetic field in the space within which the **signal-coil** is suspended. He made instruments of this kind to work both with and without electrification of the ink. Without electrification the instrument, as shown in Fig. 2, is exceedingly simple and compact, and in this form is capable of doing good work on cables of lengths up to 500 or 600 miles. When constructed for electrification of the ink, as shown in Fig. 3, it is of course available for much longer lengths of cable, but for cables such as the Atlantic cables, the original form

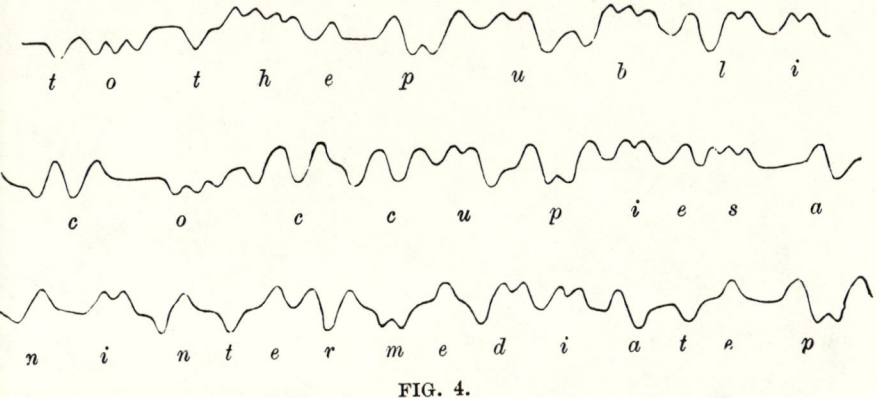

FIG. 4.

of the Siphon Recorder is that still chiefly used. The strongest magnetic field hitherto obtained by permanent magnets (of glass-hard steel) is about 3000 C. G. S. With the electromagnets of the original form of Siphon Recorder as in ordinary use a magnetic field of about or over 5000 C. G. S. is easily attained. In Fig. 4 is shown a *fac simile* of part of a message received and recorded by a Siphon Recorder, such as is shown in Fig. 1, from one of the Eastern Telegraph Co.'s Cables of about 830 miles length.

TECHNOLOGY AND SOCIETY

An Arno Press Collection

Ardrey, R[obert] L. **American Agricultural Implements.** In two parts. 1894

Arnold, Horace Lucien and Fay Leone Faurote. **Ford Methods and the Ford Shops.** 1915

Baron, Stanley [Wade]. **Brewed in America:** A History of Beer and Ale in the United States. 1962

Bathe, Greville and Dorothy. **Oliver Evans:** A Chronicle of Early American Engineering. 1935

Bendure, Zelma and Gladys Pfeiffer. **America's Fabrics:** Origin and History, Manufacture, Characteristics and Uses. 1946

Bichowsky, F. Russell. **Industrial Research.** 1942

Bigelow, Jacob. **The Useful Arts:** Considered in Connexion with the Applications of Science. 1840. Two volumes in one

Birkmire, William H. **Skeleton Construction in Buildings.** 1894

Boyd, T[homas] A[lvin]. **Professional Amateur:** The Biography of Charles Franklin Kettering. 1957

Bright, Arthur A[aron], Jr. **The Electric-Lamp Industry:** Technological Change and Economic Development from 1800 to 1947. 1949

Bruce, Alfred and Harold Sandbank. **The History of Prefabrication.** 1943

Carr, Charles C[arl]. **Alcoa, An American Enterprise.** 1952

Cooley, Mortimer E. **Scientific Blacksmith.** 1947

Davis, Charles Thomas. **The Manufacture of Paper.** 1886

Deane, Samuel. **The New-England Farmer,** or Georgical Dictionary. 1822

Dyer, Henry. **The Evolution of Industry.** 1895

Epstein, Ralph C. **The Automobile Industry:** Its Economic and Commercial Development. 1928

Ericsson, Henry. **Sixty Years a Builder:** The Autobiography of Henry Ericsson. 1942

Evans, Oliver. **The Young Mill-Wright and Miller's Guide.** 1850

Ewbank, Thomas. **A Descriptive and Historical Account of Hydraulic and Other Machines for Raising Water,** Ancient and Modern. 1842

Field, Henry M. **The Story of the Atlantic Telegraph.** 1893

Fleming, A. P. M. **Industrial Research in the United States of America.** 1917

Van Gelder, Arthur Pine and Hugo Schlatter. **History of the Explosives Industry in America.** 1927

Hall, Courtney Robert. **History of American Industrial Science.** 1954

Hungerford, Edward. **The Story of Public Utilities.** 1928

Hungerford, Edward. **The Story of the Baltimore and Ohio Railroad, 1827-1927.** 1928

Husband, Joseph. **The Story of the Pullman Car.** 1917

Ingels, Margaret. **Willis Haviland Carrier, Father of Air Conditioning.** 1952

Kingsbury, J[ohn] E. **The Telephone and Telephone Exchanges:** Their Invention and Development. 1915

Labatut, Jean and Wheaton J. Lane, eds. **Highways in Our National Life:** A Symposium. 1950

Lathrop, William G[ilbert]. **The Brass Industry in the United States.** 1926

Lesley, Robert W., John B. Lober and George S. Bartlett. **History of the Portland Cement Industry in the United States.** 1924

Marcosson, Isaac F. **Wherever Men Trade:** The Romance of the Cash Register. 1945

Miles, Henry A[dolphus]. **Lowell, As It Was, and As It Is.** 1845

Morison, George S. **The New Epoch:** As Developed by the Manufacture of Power. 1903

Olmsted, Denison. **Memoir of Eli Whitney, Esq.** 1846

Passer, Harold C. **The Electrical Manufacturers, 1875-1900.** 1953

Prescott, George B[artlett]. **Bell's Electric Speaking Telephone.** 1884

Prout, Henry G. **A Life of George Westinghouse.** 1921

Randall, Frank A. **History of the Development of Building Construction in Chicago.** 1949

Riley, John J. **A History of the American Soft Drink Industry:** Bottled Carbonated Beverages, 1807-1957. 1958

Salem, F[rederick] W[illiam]. **Beer, Its History and Its Economic Value as a National Beverage.** 1880

Smith, Edgar F. **Chemistry in America.** 1914

Steinman, D[avid] B[arnard]. **The Builders of the Bridge:** The Story of John Roebling and His Son. 1950

Taylor, F[rank] Sherwood. **A History of Industrial Chemistry.** 1957

Technological Trends and National Policy, Including the Social Implications of New Inventions. Report of the Subcommittee on Technology to the National Resources Committee. 1937

Thompson, John S. **History of Composing Machines.** 1904

Thompson, Robert Luther. **Wiring a Continent:** The History of the Telegraph Industry in the United States, 1832-1866. 1947

Tilley, Nannie May. **The Bright-Tobacco Industry, 1860-1929.** 1948

Tooker, Elva. **Nathan Trotter:** Philadelphia Merchant, 1787-1853. 1955

Turck, J. A. V. **Origin of Modern Calculating Machines.** 1921

Tyler, David Budlong. **Steam Conquers the Atlantic.** 1939

Wheeler, Gervase. **Homes for the People,** In Suburb and Country. 1855